BIRKHÄUSER

Progress in Mathematical Physics
Volume 55

The Spin

Poincaré Seminar 2007

Bertrand Duplantier
Jean-Michel Raimond
Vincent Rivasseau
Editors

Birkhäuser
Basel · Boston · Berlin

Editors:

Bertrand Duplantier
Service de Physique Théorique
Orme des Merisiers
CEA - Saclay
91191 Gif-sur-Yvette Cedex
France
e-mail: bertrand.duplantier@cea.fr

Vincent Rivasseau
Laboratoire de Physique Théorique
Université Paris-Sud
Campus d'Orsay
91405 Orsay Cedex
France
e-mail: Vincent.Rivasseau@th.u-psud.fr

Jean-Michel Raimond
Laboratoire Kastler-Brossel
Département Physique
ENS Paris
Rue Lhomond 24
75231 Paris CX 05
France
e-mail: jmr@physique.ens.fr

2000 Mathematics Subject Classification: 15A66, 32Q15, 47N50, 53C27, 81Q30, 81Txx, 81T13

Library of Congress Control Number: 2008942741

Bibliographic information published by Die Deutsche Bibliothek.
Die Deutsche Bibliothek lists this publication in the Deutsche Nationalbibliografie; detailed bibliographic data is available in the Internet at http://dnb.ddb.de

ISBN 978-3-7643-8798-3 Birkhäuser Verlag AG, Basel · Boston · Berlin

Basel · Boston · Berlin
P.O. Box 133, CH-4010 Basel, Switzerland
Part of Springer Science+Business Media
Printed on acid-free paper produced from chlorine-free pulp. TCF ∞
Printed in Germany

ISBN 978-3-7643-8798-3
9 8 7 6 5 4 3 2 1

e-ISBN 978-3-7643-8799-0
www.birkhauser.ch

CONTENTS

Foreword

This book is the eighth in a series of Proceedings for the *Séminaire Poincaré*, which is directed towards a large audience of physicists and of mathematicians.

The goal of this seminar is to provide up to date information about general topics of great interest in physics. Both the theoretical and experimental aspects are covered, with some historical background. Inspired by the Bourbaki seminar in mathematics in its organization, hence nicknamed "Bourbaphy", this Poincaré Seminar is held at the Institut Henri Poincaré in Paris, with contributions prepared in advance. Particular care is devoted to the pedagogical nature of the presentation so as to fulfill the goal of being readable by a large audience of scientists.

This new volume of the Poincaré Seminar series "The Spin" corresponds to the eleventh such Seminar, held on December 8, 2007. It describes how this once mysterious quantum reality called spin has become ubiquitous in modern physics from the most theoretical aspects down to the most practical applications of miniaturizing electronic and computer devices or helping medical diagnosis.

The first and more theoretical part of the book starts with a detailed presentation of the notion of spin by leading world expert Jürg Fröhlich. He reviews its historical development in quantum mechanics and its increasing relevance to quantum field theory and condensed matter. The next two papers discuss the exotic anyon particles. They carry a nontrivial fractional spin, hence lie in between the integer spin particles, the Bosons, and the half integer spin particles, the Fermions. The first paper by Nobel laureate Franck Wilczek gives an enticing introduction to this subject that he pioneered. The following paper by Stephane Ouvry gives a more in depth review of the corresponding mathematical formalism and its relevance in the context of the quantum Hall effect.

The second part of the book is more directly aimed at the presentation of the most advanced current experiments or applications of the notion of spin. Gerald Gabrielse reviews the extremely precise measurements of the fine structure constant and the electron magnetic moment that he and his group made by confining for months at a time a single electron in a Penning trap. This is followed by a presentation on spintronics by Nobel laureate Albert Fert and his collaborators Pierre Sénéor, Vincent Cros and Frédéric Petroff. Spintronics is the new branch of electronics in which the spin of the moving electrons is controlled and it has already lead to momentous developments in solid state physics, in particular by increasing spectacularly the capacity of magnetic memory storage devices. Finally Pierre-Jean Nacher explains the history and basic physics of magnetic resonance imaging and its application to medical diagnosis. He details the particular case of lung physiology and pathologies.

We hope that the continued publication of this series will serve the community of physicists and mathematicians at professional or graduate student level.

We thank the Commissariat à l'Énergie Atomique (Division des Sciences de la Matière) and the Daniel Iagolnitzer Foundation for sponsoring the Seminar. Special thanks are due to Chantal Delongeas for the preparation of the manuscript.

Bertrand Duplantier
Jean-Michel Raimond
Vincent Rivasseau

The Spin
B. Duplantier, J.-M. Raimond and V. Rivasseau, Eds., 1–60

Poincaré Seminar 2007

Spin, or actually: Spin and Quantum Statistics*

Jürg Fröhlich

Abstract. The history of the discovery of electron spin and the Pauli principle and the mathematics of spin and quantum statistics are reviewed. Pauli's theory of the spinning electron and some of its many applications in mathematics and physics are considered in more detail. The role of the fact that the tree-level gyromagnetic factor of the electron has the value $g_e = 2$ in an analysis of *stability* (and instability) *of matter* in arbitrary external magnetic fields is highlighted. Radiative corrections and precision measurements of g_e are reviewed. The general connection between spin and statistics, the CPT theorem and the theory of braid statistics, relevant in the theory of the quantum Hall effect, are described.

"He who is deficient in the art of selection may, by showing nothing but the truth, produce all the effects of the grossest falsehoods. It perpetually happens that one writer tells less truth than another, merely because he tells more 'truth'."
(T. Macauley, 'History', in *Essays*, Vol. 1, p 387, Sheldon, NY 1860)

Dedicated to the memory of *M. Fierz*, *R. Jost*, *L. Michel* and *V. Telegdi*, teachers, colleagues, friends.

1. Introduction to 'Spin'[1]

The 21st Century appears to witness a fairly strong decline in Society's – the public's, the politicians', the media's and the younger generations' – interest in the hard sciences, including Physics, and, in particular, in fundamental theoretical science based on precise mathematical reasoning. It is hard to imagine that reports on a discovery like the deflection of light in the gravitational field of the sun and

*Notes prepared with efficient help by K. Schnelli and E. Szabo

[1] I have to refrain from quoting literature in this introductory section – apologies!

on the underlying theory, *general relativity*, along with a photograph of its creator, *Albert Einstein*, would make it onto the front pages of major daily newspapers, as it did in 1919.

This development is, of course, not entirely accidental, and I could easily present a list of reasons for it. But let's not!

While the amount of economic wealth and added value that have been and are still being created on the basis of Physics-driven discoveries of the 19th and 20th Century is truly *gigantic*, and while one may expect that this will continue to be the case for many more years to come, fundamental physical science is confronted with a certain decline in public funding, e.g., in comparison with the life sciences. Physics is perceived to have entered a baroque state, with all the beauty that goes with it.

In this situation, it is laudable that our French colleagues are doing something to document the continuing importance and the lasting beauty of Physics: the '*Séminaire Poincaré*' (or "Bourbaphy")! I hope that the organizers of the 'Séminaire Poincaré' will find the right format and the right selection of topics for their series, and that their seminar will be accompanied by complementary activities aimed at a broader public.

This time, the topic of the 'Séminaire Poincaré' is '*Spin (and Quantum Statistics)*'. This choice of topic is not unreasonable, because, on one hand, it involves some interesting and quite fundamental experiments and theory and, on the other hand, it is connected to breathtakingly interesting and important practical applications. The scientific community sees me in the corner of mathematical physics and, thus, I have been asked to present an introductory survey of, primarily, the mathematical aspects of 'Spin and Quantum Statistics'. I am only moderately enthusiastic about my assignment, because, as I have grown older, my interests and activities have shifted more towards general theoretical physics, and, moreover, I have contributed a variety of results to, e.g., the theory of magnetism and of phase transitions accompanied by various forms of magnetic order that I cannot review, for lack of space and time.

In this short introduction, I attempt to highlight the importance of 'Spin and Quantum Statistics' for many phenomena in physics, including numerous ones that have found important technological applications, and I wish to draw attention to some of the many unsolved theoretical problems.

Our point of departure is found in the facts that electrons, positrons, neutrinos, protons and neutrons are particles with *spin* $\frac{1}{2}$ obeying *Pauli's exclusion principle*. With the exception of neutrinos, they have a *non-vanishing magnetic dipole moment*. Moreover, those particles that carry electric charge experience Coulomb- and Lorentz forces. In a magnetic field their magnetic moments and spins precess (like tops in the gravitational field of the Earth). All fundamental forces appear to be mediated by exchange of bosons of spin 1 (gauge bosons) or helicity 2 (gravitons). These facts, when exploited within the framework of quantum theory, are at the core of our theoretical description of a *vast number of phenomena* some of which we will now allude to. They are, in their majority, not very well understood, mathematically.

(1) *Chemistry.* That electrons have spin $\frac{1}{2}$ and obey the Pauli principle, i.e., are *fermions*, is one of the most crucial facts underlying all of chemistry. For example, it is the basis of our understanding of covalent bonding. If electrons were *spinless* fermions not even the simplest atoms and molecules would be the way they are in Nature: Only ortho-helium would exist, and the hydrogen molecule would not exist.

If electrons were not fermions, but bosons, there would exist ions of large negative electric charge, matter would form extremely dense clumps, and bulk matter would not be thermodynamically stable; (see section 4).

Incidentally, the hydrogen molecule is the only molecule whose stability has been deduced directly from the Schrödinger-Pauli equation with full mathematical rigour[2]. *Hund's 1st Rule* in atomic physics, which says that the total spin of the electrons in an only partially filled p-, d-, ... shell of an atom tends to be as large as possible, is poorly understood, *mathematically*, on the basis of the Schrödinger-Pauli equation.

We do not understand how *crystalline* or *quasi-crystalline order* can be derived as a consequence of *equilibrium quantum statistical mechanics.*

All this shows how little we understand about 'emergent behavior' of many-particle systems on the basis of fundamental theory. We are *not* trying to make an argument against reductionism, but one in favour of a *pragmatic attitude:* We should be reductionists whenever this attitude is adequate and productive to solve a given problem and 'emergentists' whenever this attitude promises more success!

(2) *'Nuclear and hadronic chemistry'.* At the level of fundamental theory, our understanding of binding energies, spins, magnetic moments and other properties of nuclei or of the life times of radioactive nuclei remains quite rudimentary. Presently more topical are questions concerning the 'chemistry of hadrons', such as: How far are we in understanding, on the basis of QCD, that a color-singlet bound state of three quarks (fermions with spin $\frac{1}{2}$), held together by gluons, which forms a proton or a neutron, has spin $\frac{1}{2}$? How, in the world, can we reliably calculate the magnetic dipole moments (the gyromagnetic factors) of hadrons? How far are we in truly understanding low-energy QCD? These are questions about strongly coupled, strongly correlated physical systems. They are notoriously hard to answer.

(3) *Magnetic spin-resonance.* The fact that electrons and nuclei have spin and magnetic dipole moments which can precess is at the basis of *Bloch's spin-resonance phenomenon*, which has enormously important applications in the science and technology of *imaging*; (Nobel Prizes for *Felix Bloch, Edward Purcell, Richard Ernst, Kurt Wüthrich,* ...). Of course, in this case, the basic theory is simple and well understood.

---

[2]by *G.M. Graf, J.M. Richard, M. Seifert* and myself.

(4) *Stern-Gerlach experiment*: a direct experimental observation of the spin and magnetic moment of atoms. Theory quite easy and well understood.

(5) *Spin-polarized electron emission from magnetic materials.* This is the phenomenon that when massaged with light certain magnetic materials emit *spin-polarized electrons.* It has been discovered and exploited by *Hans-Christoph Siegmann* and collaborators and has important applications in, e.g., particle physics.

(6) *Electron-spin precession in a Weiss exchange field.* When a spin-polarized electron beam is shot through a spontaneously magnetized iron-, cobalt or nickel film the spins of the electrons exhibit a *huge* precession. This effect has been discovered by H.-C. Siegmann and his collaborators and might have important applications to ultrafast *magnetic switching.* Theoretically, it can be described with the help of the Zeeman coupling of the electrons' spin to the *Weiss exchange field* (*much larger* than the magnetic field) inside the magnetized film. This effect can be interpreted as a manifestation of the $SU(2)_{\text{spin}}$-*gauge-invariance* of Pauli's electron equation; (see also section 3.3).

Related effects can presumably be exploited for the production of spin-polarized electrons and for a Stern-Gerlach type experiment for electrons.

(7) *Magnetism.* There are many materials in Nature which exhibit magnetic ordering at low temperatures or in an external magnetic field, often in combination with metallic behavior. One distinguishes between *paramagnetism, diamagnetism, ferromagnetism, ferrimagnetism, anti-ferromagnetism,* etc. In the context of the quantum Hall effect, the occurrence of *chiral spin liquids* and of chiral edge spin currents has been envisaged; ....

The theory of paramagnetism is due to Pauli; it is easy. The theoretical basis of diamagnetism is clear. The theory of *anti-ferromagnetism* and *Néel order* at low temperatures in insulators is relatively far advanced. But the theory of ferromagnetism and the appearance of spontaneous magnetization is disastrously *poorly understood, mathematically.* Generally speaking, it is understood that spontaneous (ferro- or anti-ferro-) magnetic order arises, at low enough temperature, by a conspiracy of electron spin, the Pauli principle and Coulomb repulsion among electrons. The earliest phenomenological description of phase transitions accompanied by the appearance of magnetic order goes back to *Curie* and *Weiss. Heisenberg* proposed a quantum-mechanical model inspired by the idea of direct electron exchange interactions between neighboring magnetic ions (e.g. Fe) in a crystalline back ground. While it has been shown, mathematically, that the classical Heisenberg model (large-spin limit) and the *Heisenberg anti-ferromagnet* exhibit the expected phase transitions[3], *no* precise understanding of the phase transition in the *Heisenberg ferromagnet* (finite spin) has been achieved, yet.

Most of the time, the microscopic origin of *exchange interactions* between spins in magnetic materials remains poorly understood, mathematically. No mathematically precise understanding of ferromagnetic order in models of *itinerant*

[3]in work by *Simon, Spencer* and myself, and by *Dyson, Lieb* and Simon; and followers.

electrons, such as the weakly filled one-band *Hubbard model*, has been reached, yet. However, there is some understanding of Néel order in the half-filled one-band Hubbard model (‘*Anderson mechanism*’) and of ferromagnetic order in Kondo lattice models with a weakly filled conduction band (*Zener's mechanism* of *indirect exchange*), which is mathematically rather precise at *zero* temperature.

Realistic *spin glasses* are extremely poorly understood, theory-wise.

Altogether, a general theory of magnetism founded on basic equilibrium quantum statistical mechanics still remains to be constructed!

Of course, magnetism has numerous applications of great importance in magnetic data storage, used in computer memories, magnetic tapes and disks, etc.

(8) *Giant and colossal magneto-resistance.* The discoverers of giant magneto-resistance, *Albert Fert* and *Peter Grünberg*, have just been awarded the 2007 Nobel Prize in Physics. Their discovery has had fantastic applications in the area of data storage and -retrieval. It will be described at this seminar by Fert and collaborators. Suffice it to say that electron spin and the electron's magnetic moment are among the main characters in this story, and that heuristic, but quite compelling theoretical understanding of these phenomena is quite advanced.

(9) *Spintronics.* This is about the use of electron spin and multi-spin entanglement for the purposes of quantum information processing and quantum computing. Presently, it is a hot topic in mesoscopic physics. Among its aims might be the construction of scalable arrays of interacting quantum dots (filled with only few electrons) for the purposes of quantum computations; (the spins of the electrons would store the Qbits).

(10) *The rôle of electron spin and the Weiss exchange field in electron – or hole – pairing mechanisms at work in layered high-temperature superconductors.* This is the idea that the Weiss exchange field in a magnetic material can produce a strong attractive force between two holes or electrons (introduced by doping) in a spin-singlet state, leading to the formation of Schafroth pairs, which, after condensation, render such materials superconducting.

(11) *The rôle played by spin and by particle-pairing in the miraculous phase diagram of* ^{3}He *and in its theoretical understanding.* The rôle played by spin in the physics of ‘heavy fermions’.

(12) *The rôle of the Pauli principle (and spin, in particular neutron spin) in the physics of stars.* The theory of the Chandrasekhar limit for white dwarfs and neutron stars is based on exploiting the Pauli principle for electrons or neutrons in an important way. The superfluidity expected to be present in the shell of a neutron star is a phenomenon intimately related to the spin of the neutron, neutron pairing and pair condensation.

Many of these topics have been close to my heart, over the years, and I have written hundreds of pages of scientific articles that have been read by only few people. One could easily offer a one-year course on these matters. But, in the following sections, I really have to focus on just a few *basic* aspects of 'Spin and Quantum Statistics'.

Acknowledgments. I thank C. Bachas, B. Duplantier and V. Rivasseau for inviting me to present a lecture at the 'Séminaire Poincaré' and my teachers and numerous collaborators for all they have taught me about 'Spin and Quantum Statistics', over many years. I am very grateful to K. Schnelli for his help.

Remark. These notes have been written at a 'superluminal' speed and are therefore likely to contain errors and weaknesses, which I wish to offer my apologies for.

2. The Discovery of Spin and of Pauli's Exclusion Principle, Historically Speaking

My main sources for this section are [1–6]. Let us dive into a little history of science, right away.

2.1. Zeeman, Thomson and others, and the discovery of the electron

Fairly shortly before his death, in 1867, *Michael Faraday* made experiments on the influence of 'strong' magnetic fields on the frequency of light emitted by excited atoms or molecules. He did this work in 1862 and did not find any positive evidence for such an influence. In the 1880's, the American physicist *Henry Augustus Rowland* invented the famous 'Rowland gratings', which brought forward much higher precision in measuring wave lengths of spectral lines.

In 1896, *Pieter Zeeman*, a student of *Kamerlingh Onnes* and *Hendrik Antoon Lorentz*, took up Faraday's last experiments again, using Rowland gratings. He found that the two sodium D-lines are broadened when the magnetic field of an electromagnet[4] is turned on. He proposed to interpret the effect in terms of Lorentz' theory of charges and currents carried by fundamental, point-like particles. In 1895, Lorentz had introduced the famous *Lorentz force* acting on charged particles moving through an electromagnetic field. When Zeeman had discovered the effect named after him Lorentz proposed a model of harmonically bound charged particles of charge e. When a magnetic field $\vec{H}$ is turned on in a direction perpendicular to the plane of motion of such a particle the angular frequency of its motion changes by the amount

$$\Delta\omega = \frac{e}{mc}|\vec{H}|\,,$$

where m is its mass and c is the speed of light. Using Lorentz' formula, Zeeman inferred from the broadening of the sodium lines that

$$\frac{e}{m} \simeq 10^7 \text{emu/g} \qquad (1.76 \times 10^7 \text{emu/g})\,.$$

[4]Concerning electromagnets, one could embark on a report of the important contributions and inventions of *Pierre Weiss*, once upon a time a professor at ETH Zurich.

In 1897, Zeeman discovered a *splitting* of the blue line of cadmium, in rough agreement with Lorentz' theoretical expectations. From polarization effects he inferred that e is *negative*. *George Stoney* had earlier provided an estimate for the elementary electric charge e. Thus, Zeeman could have predicted the mass of the charged particle that emits electromagnetic radiation from the 'interior' of an atom or molecule, the *electron*.

In the same year, the quotient $\frac{e}{m}$ was measured in experiments with *cathode rays*, first by *Emil Wiechert*[5], who conjectured that such rays consist of charged particles with a very small mass m (=mass of an electron); then - with very high accuracy - by *Walter Kaufman* and, more or less simultaneously, by *Joseph J. Thomson*, who also proposed Wiechert's charged-particle picture. In 1899, Thomson measured the value of e by cloud chamber experiments, and, in 1894, he had obtained some bounds on the speed of propagation of cathode rays, showing that this speed is considerably smaller than the speed of light. This combination of accomplishments led to the common view that J.J. Thomson is the discoverer of the *electron*.

After the discovery of relativistic kinematics in 1905, by *Einstein*, experiments with electrons became the leading tool to verify the kinematical predictions of the *special theory of relativity*.

2.2. Atomic spectra

"Spectra are unambiguous visiting cards for the gases which emit them."
(Abraham Pais [1])

Spectroscopy started in Heidelberg with the work of *Gustav Kirchhoff* (1859) and *Robert Bunsen*. Against the philosophical prejudices of *Auguste Comte*, Kirchhoff concluded with the help of *absorption spectroscopy* that the solar atmosphere must contain sodium[6]. Kirchhoff and Bunsen are the fathers of modern optical spectroscopy and its application as an exploratory tool.

The first three lines of the *hydrogen spectrum* were first observed by *Julius Plücker* in 1859, then, more precisely, by *Anders Ångström* in 1868. Searches for patterns in spectral lines started in the late 1860's. The first success came with *Stoney* in 1871. The break-through was a famous formula,

$$\lambda_n = \frac{Cn^2}{n^2 - 4},$$

where the λ_n are wave lengths of light emitted by hydrogen, C is some constant, and $n = 3, 4, \ldots$, discovered by *Johann Jakob Balmer* in 1885. In 1892, *Carl Runge* and *Heinrich Kayser* made precise measurements of spectral lines of 22 elements. *Runge* and *Friedrich Paschen* discovered the spectra of ortho- and parahelium. A precursor of the *Rydberg-Ritz combination principle* was discovered in 1889 by *Johannes Rydberg*, its general form was found by *Walther Ritz* in 1908.

---

[5] Of fame also in connection with the Liénard-Wiechert potentials.

[6] "It's not philosophy we are after, but the behaviour of real things." (Richard P. Feynman)

Precursors of *Rutherford's planetary model* of the atom (1911) can be found in remarks by *Heinrich Hertz* (lectures about the constitution of matter in Kiel), *Hermann von Helmholtz*, *Jean Perrin* (1901), *Hantaro Nagaoka* (1903), and *J.J. Thomson* (1906).

In 1913, *Niels Bohr* came up with his quantum theory of the hydrogen atom[7], with the idea that atomic spectra arise by photon emission during transitions of an electron from one 'stationary state' (a term introduced by Bohr) to another, and with the *Bohr frequency condition*, which has a precursor in *Einstein's* work of 1906 on Planck's law for black-body radiation. Bohr's results provided a quantum-theoretical 'explanation' of Balmer's formula and of a special case of the Rydberg-Ritz combination principle.

Subsequent to Bohr's discoveries, in attempts to interpret the so-called 'fine structure' of atomic spectra discovered by *Albert Michelson* (1892) and Paschen (1915), Bohr's quantum theory was to be married with the special theory of relativity. The pioneer was *Arnold Sommerfeld* (1916). He introduced the *fine structure constant*

$$\alpha = \frac{e^2}{\hbar c}.$$

Sommerfeld's formula for the relativistic hydrogen energy spectrum is

$$E_{n,l} = -\mathrm{Ry}\left[\frac{1}{n^2} + \frac{\alpha^2}{n^3}\left(\frac{1}{l+1} - \frac{3}{4n}\right)\right] + \mathcal{O}(\alpha^4), \tag{2.1}$$

where $n = 1, 2, 3, \ldots, l = 0, 1, \ldots, n-1$ and Ry is the Rydberg constant. Of course $\vec{L}$, with $|\vec{L}| \simeq \hbar(l+1)$, is the (quantized) angular momentum of the electron orbiting the nucleus.

In trying to explain experimental results of Paschen, *Bohr* and, independently, *Wojciech Rubinowicz* (a collaborator of Sommerfeld) found the *selection rule*

$$\Delta l = \pm 1 \tag{2.2}$$

for transitions between stationary states.

This rule did not work perfectly. In 1925, in their first publication and after ground-breaking work of *Wolfgang Pauli*, *George Uhlenbeck* and *Samuel Goudsmit* proposed a modification of the Bohr-Rubinowicz selection rule: In (2.1), write

$$l + 1 = j + \frac{1}{2}, \tag{2.3}$$

with j half-integer, and replace (2.2) by

$$\Delta j = 0, \pm 1. \tag{2.4}$$

[7] His theory has a more incomplete precursor in the work of *Arthur Erich Haas* (1910).

This reproduced data for the fine structure of the He^+ spectrum perfectly. Here, the half-integer quantum number j appears. Similar ideas were proposed independently by *John Slater.*

Of course, the half-integer nature of j (for atoms or ions with an odd number of bound electrons) is related to *electron spin*; as everybody knows nowadays. Actually, half-integer quantum numbers were first introduced systematically by *Alfred Landé* in an analysis of the Zeeman effect and correctly interpreted, by Pauli, as "due to a peculiar classically not describable two-valuedness of the quantum theoretical properties of the valence electron", in 1924.

We have now reached the period when *electron spin* enters the scene of physics. I shall briefly sketch how it was discovered by Pauli towards the end of 1924.

2.3. Pauli's discovery of electron spin and of the exclusion principle

Pauli's papers on electron spin and the exclusion principle are [7,8,11]. In [7], he analyzes what is known as the *'anomalous Zeeman effect'*, namely the Zeeman effect in *weak* magnetic fields (when relativistic spin-orbit terms dominate over the Zeeman term in the atomic Hamiltonian). This theme is taken up again in [8,11] and leads him to discover electron spin and the exclusion principle. Let us see how this happened!

In [7], Pauli started from the following facts and/or assumptions; (I follow modern notation and conventions).

(1) Spectral terms (energies corresponding to stationary states) can be labeled by *'quantum numbers'*:
   - (i) A principal quantum number, n, (labeling shells).
   - (ii) $L = 0, 1, 2, 3, \dots (S, P, D, F, \dots)$ with $L < n$ – our orbital angular momentum quantum number – and $M_L = -L, -L+1, \dots, L$ – the magnetic quantum number.
   - (iii) $S = 0, 1/2, 1, \dots$, and $M_S = -S, -S+1, \dots, S$.
   - (iv) The terms of a multiplet with given L and S are labeled by a quantum number J (our total angular momentum quantum number), whose possible values are $J = L+S, L+S-1, \dots, |L-S|$, and a magnetic quantum number $M = -J, -J+1, \dots, J$.

(2) There are *selection rules* for the allowed transitions between stationary states:
$$\Delta L = \pm 1,\ \Delta S = 0,\ \Delta J = 0, \pm 1 \text{ (with } J = 0 \to J = 0 \text{ forbidden).}$$

(3) Denoting by Z the atomic number of a *neutral* atom, one has the *correspondence*
$$\begin{aligned} Z \quad \text{even} \quad &\longleftrightarrow \quad S,\, J \text{ integer}\,, \\ Z \quad \text{odd} \quad &\longleftrightarrow \quad S,\, J \text{ half-integer}\,. \end{aligned}$$

(4) *Bohr's frequency condition* (i.e., the formula for the frequency of light emitted in a transition from one stationary state to a lower-lying one.)

(5) *Line splittings in a magnetic field* $\vec{H}$. If Zeeman splitting dominates fine structure splitting (*Paschen-Back effect*) then the energy splitting is given by

$$\Delta E \simeq (M_L + 2M_S)\mu_0|\vec{H}|\,, \tag{2.5}$$

where $\mu_0 = \frac{e\hbar}{2mc}$ is *Bohr's magneton* (actually introduced by Pauli in 1920).

If fine structure (spin-orbit interactions) dominates over Zeeman splitting (*anomalous Zeeman effect*) a term with quantum number J splits into $2J+1$ equidistant levels labeled by a 'magnetic quantum number' $M = -J, -J+1, \dots, J$, and the energy splitting for a term with given L, S, J and M is given by

$$\Delta E = Mg\mu_0|\vec{H}|\,,$$

where g is the *Landé factor*,

$$g = \frac{3}{2} + \frac{S(S+1) - L(L+1)}{2J(J+1)}\,. \tag{2.6}$$

The selection rules for transitions are given by

$$\Delta M = 0, \pm 1\,.$$

Starting from the Paschen-Back effect, Pauli postulates that the sum of energy levels in a multiplet with given L and M is a *linear* function of $|\vec{H}|$ when one passes from strong to weak magnetic fields. He then determines Landé's g-factors uniquely from the energy splittings in large fields and the 'sum rule' just stated. Nowadays, these calculations are an elementary exercise in the algebra of quantum-mechanical angular momenta (see, e.g., [6]), which I will not reproduce. Pauli concludes his paper [7] with prophetic remarks that a derivation of the 'laws' he analyzed within the principles of the (old) quantum theory then known, does not appear to be possible; that the connection between angular momentum and magnetic moment predicted by Larmor's theorem does not generally hold ($g_e = 2\,!$); and that the appearance of half-integer values of M and J goes beyond the quantum theory of quasi-periodic mechanical systems.

Soon afterwards, Pauli started to think about the problem of completion of electron shells in atoms and the doublet structure of alkali spectra. This led him to his important paper [8]. Before I sketch the contents of [8], I recall a standard calculation of the gyromagnetic ratio between magnetic moment $\vec{M}$, and angular momentum $\vec{L}$. We consider a distribution of rotating, charged, massive matter. If we assume that the charge and mass densities are proportional to each other then

$$\frac{|\vec{M}|}{|\vec{L}|} = \frac{|q|}{2mc}\,, \tag{2.7}$$

where q is the total charge and m the total mass. Apparently, the Landé factor is $g = 1$. If the same calculation is done using relativistic kinematics (as Pauli did in [8]) one finds that

$$\frac{|\vec{M}|}{|\vec{L}|} = \frac{|q|}{2mc} \cdot (\overline{\gamma})^{-1}, \tag{2.8}$$

where $\gamma = (1 - \frac{v^2}{c^2})^{-1/2}$, v is the speed of a mass element, and $\overline{(\cdot)}$ denotes a suitable average. Note that $(\overline{\gamma})^{-1} < 1$!

When Pauli worked on paper [8] the prejudice was that, for alkaline metals, the quantum number S was related to the angular momentum of the *core* (filled shells) of an atom. It was then to be expected that it would correspond to a magnetic moment $\vec{M}$ with

$$|\vec{M}| = \frac{e}{2mc}(\overline{\gamma})^{-1} S.$$

Thus, the Landé factor of the core should have come out to be

$$g_{\text{core}} = (\overline{\gamma})^{-1} < 1. \tag{2.9}$$

Since the electrons in the core of large-Z elements are relativistic, the prediction of the 'core model' would have been that g_{core} is *measurably smaller* than 1.

However, formula (2.5), well confirmed, for large $|\vec{H}|$, in experiments by Runge, Paschen and Back for large-Z elements, and Landé's formula (2.6) were only compatible with

$$g_{\text{core}} = 2.$$

Pauli concluded that S could not have anything to do with the angular momentum of the core (filled shells) of an atom. He goes on to propose that filled shells have angular momentum 0 and do *not* contribute to the magnetic moment of the atom. By studying experimental data for the Zeeman effect in alkali atoms, he arrives at the following *key conclusion*:

> "The closed electron configurations shall not contribute to the magnetic moment and angular momentum of the atom. In particular, for the alkalis, the angular momenta of, and energy changes suffered by, the atom in an external magnetic field shall be regarded exclusively as an effect of the valence electron ('Leuchtelektron'), which is also the source of the magneto-mechanical anomaly[8]. The doublet structure of the alkali spectra, as well as the violation of the Larmor theorem are, according to this point of view, a result of a classically not describable two-valuedness of the quantum-theoretical properties of the valence electron."

Thus, Pauli had discovered the *spin of the electron* and the *'anomaly'* in its g-factor, $g_e = 2$. (See [9] for a recent study why $g = 2$ is the natural value of the tree-level gyromagnetic ratio of charged elementary particles.)

Soon, *Ralph Kronig* and, independently, *Uhlenbeck* and *Goudsmit* interpreted the quantum number S as due to an intrinsic rotation of electrons, picturing them as little charged balls. Kronig explained his idea to Pauli, who thought

[8] $g_e = 2$!

it was nonsense[9], and Kronig did not publish it. Uhlenbeck and Goudsmit were confronted with objections by *Lorentz* against their idea related to the fact that $g_e = 2$, and wanted to withdraw their paper from publication, but *Ehrenfest* convinced them to go ahead and publish it.

Now comes the problem of the *Thomas precession*: As had been discovered by Einstein and explained by him to his colleagues working on the quantum theory, an electron traveling through an electric field $\vec{E}$ with a velocity $\vec{v}$ feels a magnetic field

$$\vec{B}' = -\frac{\vec{v}}{c} \wedge \vec{E} + \mathcal{O}\left(\frac{v^2}{c^2}|\vec{E}|\right) \tag{2.10}$$

in its rest frame. If its magnetic moment in the rest frame is denoted by $\vec{M}$ one expects that its spin $\vec{S}$, will exhibit a precession described, in its rest frame, by

$$\frac{d\vec{S}}{dt} = \vec{M} \wedge \vec{B}', \tag{2.11}$$

corresponding to a magnetic energy

$$U' = -\vec{M} \cdot \vec{B}'. \tag{2.12}$$

For an electron in the Coulomb field of a nucleus

$$e\vec{E} = -\frac{\vec{x}}{r}\frac{\mathrm{d}V(r)}{\mathrm{d}r}, \tag{2.13}$$

where r is the distance to the nucleus, and V is the Coulomb potential. Plugging (2.13) into (2.10) and (2.10) into (2.12), we find that

$$U' = \frac{g_e}{2(mc)^2}(\vec{S} \cdot \vec{L})\frac{1}{r}\frac{\mathrm{d}V(r)}{\mathrm{d}r},$$

where $\vec{L}$ is the orbital angular momentum, the well-known spin-orbit interaction term. If this formula is taken literally and compared with Sommerfeld's calculation of the fine structure (see Eq. (2.1)) one finds that g_e must be 1. This is a contradiction to the value $g_e = 2$ found in the analysis of the Zeeman effect for alkali atoms.

This contradiction vexed many people, foremost Pauli, and *Heisenberg* communicated it to Uhlenbeck and Goudsmit when he saw their paper, ("ihre mutige Note"). It was resolved by *Llewellyn Thomas*, in February 1926. Thomas pointed out that the rest frame of an electron moving in the Coulomb field of a nucleus is actually *rotating* relative to the laboratory frame. The angular velocity of that rotation is denoted by $\vec{\omega}_T$. Then the equation for the precession of the electron's spin in a *non-rotating* frame moving with the electron is given by

---

[9]One might say: *correctly*, (since $s = \frac{1}{2}$ is far away from the classical limit $s = \infty$).

$$\left(\frac{\mathrm{d}\vec{S}}{\mathrm{d}t}\right)_{\text{non-rotating}} = \left(\frac{\mathrm{d}\vec{S}}{\mathrm{d}t}\right)_{\text{rest frame}} + \vec{\omega}_T \wedge \vec{S}\,, \tag{2.14}$$

with $\left(\frac{\mathrm{d}\vec{S}}{\mathrm{d}t}\right)_{\text{rest frame}}$ given by (2.11). The 'magnetic energy' in the non-rotating frame is then given by

$$U = U' + \vec{S} \cdot \vec{\omega}_T\,. \tag{2.15}$$

The problem now boils down to calculating $\vec{\omega}_T$. This is an exercise in composing Lorentz boosts whose solution can be looked up, e.g., in [10]. The formula for $\vec{\omega}_T$ is

$$\omega_T = \frac{1}{2}\frac{\vec{a} \wedge \vec{v}}{c^2}\left(1 + \mathcal{O}\left(\frac{v^2}{c^2}\right)\right), \tag{2.16}$$

where $\vec{a}$ is the acceleration of the electron, which, in an electric field, is given by $-\frac{e}{m}\vec{E}$, up to corrections $\mathcal{O}\left(\frac{v}{c}\right)$. Then U is given by

$$U \simeq \frac{(g_{\mathrm{e}} - 1)e}{2mc}\vec{S} \cdot \left(\frac{\vec{v}}{c} \wedge \vec{E}\right), \tag{2.17}$$

which, in the Coulomb field of a nucleus, becomes

$$U \simeq \frac{(g_{\mathrm{e}} - 1)e}{2(mc)^2}\vec{S} \cdot \vec{L}\frac{1}{r}\frac{\mathrm{d}V}{\mathrm{d}r}\,. \tag{2.18}$$

This expression reproduces the correct fine structure. Expression (2.16) for the Thomas precession frequency and the second term on the R.S. of (2.15) have been verified, experimentally, in data for spectra of nuclei (where the Landé g-factor does *not* take the value $g = 2$).

Thomas' observations convinced people, including Einstein and Pauli, and boosted the acceptance of the naive interpretation of electron spin proposed by Uhlenbeck and Goudsmit in the physics community.

I conclude my excursion into the history of the discovery of spin with comments on precursors.

In 1900, *George Francis FitzGerald* had raised the question whether magnetism might be due to a rotation of electrons. In 1921, *Arthur Compton* proposed that "it is the electron rotating about its axis which is responsible for ferromagnetism"; (see [1], page 279). The same idea was proposed by *Kennard*, who also argued (independently of *Abraham*), that g_{e} could have the value 2. In 1924 (before he wrote the papers [8] and [11]), Pauli proposed that the atomic nucleus must, in general, have a non-vanishing angular momentum, which was relevant for an explanation of hyperfine splitting. (Whether his idea influenced Uhlenbeck and Goudsmit, or not, is unclear but rather unlikely.) Independently of (and priorly to) Uhlenbeck and Goudsmit, *Kronig* and *Urey* anticipated their idea, and *Bose* had the idea that photons carry an intrinsic 'spin' (or helicity, as we would call it nowadays).

Almost all these ideas were somewhat flawed or incomplete. For example, we understand – since Heisenberg's proposal of a model of ferromagnetism – that the *Pauli principle* plays as important a rôle in explaining ferromagnetism as electron spin.

Thus, let me briefly recall the history of the discovery of *Pauli's exclusion principle*[10]. This discovery was made on the basis of Bohr's work on the periodic table of elements, in particular his *'permanence principle'* (electrons in the shell of an ion keep their quantum numbers when further electrons are added), and of an important paper by *Edmund Stoner* [12]. Stoner classified electron configurations corresponding to given values of the quantum numbers L and J and found, for alkali atoms, that the *total* number of electrons in such a configuration is identical to the number of terms in the Zeeman spectrum of these atoms, namely $2(2L+1)$, for every $L < n$ (=principal quantum number). Pauli accidentally came across Stoner's paper. Considering alkali spectra, Pauli notices that "the number of states in a magnetic field for given values of L and J is $2J + 1$, the number of states for both doublets together, with L fixed, is $2(2L + 1)$". Using Bohr's permanence principle, he extends his counting of states to more complicated atoms and to *all* electrons in the hull of an atom. He concludes that "every electron in an atom can be characterized by a principal quantum number n and three additional quantum numbers (L, J, m_J)", (with $J = L \pm \frac{1}{2}$). He notices that, for $L = 0$, there are four possible states for two electrons with different principal quantum numbers, but only one when their principal quantum numbers agree. He then goes on to explain Stoner's and his observations by postulating that each state characterized by quantum numbers (n, L, J, m_J) can be occupied by at most *one* electron. (Pauli had actually defined $L, J = L \pm \frac{1}{2}$, and $m_J = J, J-1, \ldots, -J$ for *single* electrons.) This is the *exclusion principle.* Pauli concludes his paper with the sentence:

> "The problem of a more coherent justification of the general rules concerning equivalent electrons in an atom here proposed can probably only be attacked successfully after a further deepening of the fundamental principles of quantum theory."

Further deepening of the fundamental principles of quantum theory was to come forward, just a few months later, starting with the work of *Heisenberg* [13], followed by a paper by *Max Born* and *Pascual Jordan* [14], the "Drei-Männer-Arbeit" [15], Dirac's first contributions to the new matrix mechanics [16] (published before he earned his PhD degree under *Fowler* in 1926), and, finally, by *Schrödinger's* work on wave mechanics, in 1926; see [17]. When Heisenberg started to do his fundamental work resulting in the paper [13], his friend Pauli was momentarily fed up with quantum theory and worked on Kaluza-Klein theory.

---

[10] A name introduced by *Dirac* in 1925.

The quantum mechanics of angular momentum, including half-integer angular momentum, was fully developed in [15]. Pauli's exclusion principle was reformulated, quantum mechanically, as saying that many-electron states (wave functions) must be *totally anti-symmetric* under permutations of the positions and spins of individual electrons. An early contribution in this direction was in a paper by Heisenberg, the general formulation is due to Dirac (1926) and, in its definitive version, to *Eugene Wigner* (1928), who profited from his friend's, *John von Neumann*, knowledge of the permutation groups and their representations. The first applications to statistical mechanics were made by *Jordan*[11], *Fermi* and Dirac, in 1926, (Fermi-Dirac statistics).

Bose-Einstein statistics (for particles with integer spin) was introduced by *Bose* (for photons) and *Einstein* (for ideal monatomic quantum gases) in 1924. Its quantum-mechanical reformulation says that wave functions of many *identical* bosons must be totally symmetric under permutations of these particles. Einstein predicted *Bose-Einstein condensation* for non-relativistic Bose gases (and used a *wave picture* for the atoms in the gas) in 1924.

It should be added that the spin and the value $g_{\mathrm{e}} = 2$ of the gyromagnetic factor of the electron, as well as the fine structure of the hydrogen spectrum that led to the discovery of the Thomas precession, all found a natural explanation when *Dirac* discovered his relativistic electron equation named after him, in 1927; see [18]. We will briefly return to this equation, later.

I will now leave the history of the discoveries of spin and quantum statistics and proceed to sketching some highlights, mathematical and physical ones, that emerged from these discoveries, *not* attempting to provide a historical perspective and jumping over many important developments. I try to provide a glimpse at the usefulness of Mathematics in formulating and understanding the laws of Physics.

3. Some of the Mathematics of Spin and a Theorem of Weyl[12]

The model of space and time underlying non-relativistic quantum mechanics is inherited from *Newtonian mechanics:* Physical space is homogeneous and isotropic, and an appropriate model is three-dimensional Euclidian space $\mathbb{E}^3$. Time is modelled by the real line, with the standard order relation and metric. Space-time $\mathcal{N}$ is given by $\mathbb{E}^3 \times \mathbb{R}$. Events are identified with points in $\mathcal{N}$. The time difference between two events and the spatial distance between them are invariants. Dynamical symmetries of autonomous physical systems are described by the group of Euclidian motions, boosts and time translations, the so-called *Galilei group.*

The model of space-time underlying special-relativistic quantum theory (gravity neglected) is usually taken to be the one proposed by *Poincaré* and *Minkowski.*

---

[11] Jordan was apparently first in discovering Fermi-Dirac statistics. But the editor of 'Zeitschrift für Physik', Max Born, forgot to send Jordan's paper to the publisher during his stay in America. I thank N. Straumann for communicating this to me.

[12] Sources for the material in this section are [6, 19–24].

Space-time is denoted by $\mathcal{N} \simeq \mathbb{R}^4$, events are labeled by points in $\mathcal{N}$, and the only invariant for a pair of events labeled by the points $(t, \vec{x})$ and $(t', \vec{x}')$ is given by

$$c^2(t-t')^2 - |\vec{x} - \vec{x}'|^2 ,$$

where c is the speed of light. If this quantity is positive then $\operatorname{sign}(t - t')$ is an invariant, too. Symmetries of autonomous physical systems are described by the Poincaré transformations of $\mathcal{N}$, which form the *Poincaré group.*

The Galilei group is recovered from the Poincaré group by 'group contraction', as the 'deformation parameter' $1/c$ tends to 0. As long as recoil on the gravitational field is neglected and this field is treated as an *external field,* there are many good models of Lorentzian space-times that can serve as receptacles for a quantum theory. But a good model of space-time underlying a quantum theory of matter and gravitation is not known, yet!

What has all this got to do with spin? Both the Galilei and the Poincaré group in $d = n+1$ dimensions (with $n = 3$, in nature) contain the group $\mathrm{SO}(n)$ of spatial rotations as a subgroup: Generally speaking, if physical space is isotropic spatial rotations are dynamical symmetries of autonomous non-relativistic and special relativistic quantum-mechanical systems, and we must ask how these symmetries are represented on the space of states of such a system, and what this has got to do with spin.

Let G be any group of symmetries of a quantum-mechanical system with a Hilbert space $\mathscr{H}$ of pure state vectors. *Eugene Wigner* has shown that symmetry transformations labeled by elements of G are represented as unitary or anti-unitary operators acting on $\mathscr{H}$, and that these operators must define a *projective representation* of G on $\mathscr{H}$, (because the phase of a vector in $\mathscr{H}$ is not observable; the space of *pure states* being given by projective space over $\mathscr{H}$). *Valentin Bargmann* has shown that if G is a *connected, compact* Lie group then all projective representations of G are given by unitary representations of the *universal covering group* $\tilde{G}$ associated with G.

If $G = \mathrm{SO}(n), n = 2, 3, 4, \ldots,$ (the rotation group in n dimensions), then

$$\widetilde{G} = \begin{cases} \mathbb{R}\,, & n = 2 \\ \mathrm{SU}(2)\,, & n = 3 \\ \mathrm{Spin}(n)\,, & n \text{ general}\,. \end{cases}$$

The *spin* of a quantum-mechanical particle is viewed as its intrinsic angular momentum and is thus described in terms of the generators of rotations in an irreducible, unitary representation of the quantum-mechanical rotation group $\mathrm{Spin}(n)$, where n is the dimension of physical space. For $n = 2$, these representations are given by the characters of the group $\mathbb{R}$, i.e., labeled by a *real number* s, called the 'spin of the representation'. For $n = 3$, the representation theory of (the Lie algebra of) $\mathrm{Spin}(3) = \mathrm{SU}(2)$ has been worked out in [15] and is taught in every course on introductory quantum mechanics. Irreducible representations are labeled by their 'spin' $s = 0, \frac{1}{2}, 1, \frac{3}{2}, \ldots$. For general n, we refer, e.g., to [24].

We do not have to go into this theory in any detail. We just note that, for $n \geq 3$, Spin(n) is a two-fold cover of SO(n) and that, as a consequence, there are irreducible representations of Spin(n) that are *single-valued representations* of SO(n) (rotations through an angle 2π =identity) labeled by '$\sigma = 1$', and representations of Spin(n) that are '*double-valued representations*' of SO(n) (rotations through an angle $2\pi = -$identity) labeled by '$\sigma = -1$'.

For an understanding of differential-geometric aspects of 'spin' it is useful to consider the quantum mechanics of a single non-relativistic particle with spin moving in a physical space described by a rather general n-dimensional manifold. Of course we are mainly interested in the examples $n = 2$ (planar physics) and $n = 3$; but, for purposes of applications in mathematics, it pays to be a little general, here. We are interested in formulating non-relativistic quantum mechanics on a space-time $\mathcal{N}$ of the form

$$\mathcal{N} = \mathcal{M} \times \mathbb{R},$$

where physical space, $\mathcal{M}$, is a general smooth, orientable spin$^{\mathbb{C}}$ manifold, equipped with a Riemannian metric g, and $\mathbb{R}$ denotes time. Our goal is to derive Pauli's wave equation for a non-relativistic electron with spin moving in $\mathcal{M}$ under the influence of an external electromagnetic field and to also consider the quantum mechanics of positronium (a bound electron-positron pair). For the standard choice $\mathcal{M} = \mathbb{E}^3$ of direct interest in physics, Pauli's wave equation was discovered in [19].

3.1. Clifford algebras and spin groups

Let $\mathcal{F}_k$ be the unital *algebra generated by elements $b^1, \ldots, b^k$ and their adjoints $b^{1*}, \ldots,$
b^{k*} satisfying the canonical anti-commutation relations (CAR)

$$\{b^i, b^j\} = \{b^{i*}, b^{j*}\} = 0, \qquad \{b^i, b^{j*}\} = \delta^{ij}, \tag{3.1}$$

where $\{A, B\} := AB + BA$. The algebra $\mathcal{F}_k$ has a unique (up to unitary equivalence) irreducible unitary representation on the Hilbert space $S := \mathbb{C}^{2^k}$ given by

$$\begin{aligned} b^j &= \tau_3 \otimes \cdots \otimes \tau_3 \otimes \tau_- \otimes \mathbb{1}_2 \otimes \cdots \otimes \mathbb{1}_2, \\ b^{j*} &= \tau_3 \otimes \cdots \otimes \tau_3 \otimes \tau_+ \otimes \mathbb{1}_2 \otimes \cdots \otimes \mathbb{1}_2, \end{aligned} \tag{3.2}$$

with $\tau_\pm := \frac{1}{2}(\tau_1 \pm \mathrm{i}\tau_2)$ in the jth factor; τ_1, τ_2 and τ_3 are the usual 2×2 Pauli matrices. The representation (3.2) is faithful, and hence $\mathcal{F}_k \simeq M(2^k, \mathbb{C})$, the algebra of $2^k \times 2^k$ matrices over the complex numbers.

Let V be a real, oriented, n-dimensional vector space with scalar product $\langle \cdot, \cdot \rangle$. The complexified Clifford algebra Cl(V) is the algebra generated by vectors $c(v), c(w)$, linear in v, w, with v and w in $V \otimes \mathbb{C}$, subject to the relations

$$\{c(v), c(w)\} = -2\langle v, w \rangle. \tag{3.3}$$

If $e^1, \dots, e^n$ is an orthonormal basis of V, $n = \dim V$, then (3.3) implies that

$$\{c(e^i), c(e^j)\} = -2\delta^{ij} .$$

A *operation is defined by

$$c(v)^* = -c(\bar{v}) , \tag{3.4}$$

$v \in V \otimes \mathbb{C}$. Let $n = 2k + p$, where $p = 0$ or 1 is the parity of n. Setting

$$\begin{aligned} c(e^{2j-1}) &:= b^j - b^{j*} , \\ c(e^{2j}) &:= \mathrm{i}\left(b^j + b^{j*}\right) , \end{aligned} \tag{3.5}$$

$j = 1, \dots, k$, and, for $p = 1$,

$$c(e^n) := \pm \mathrm{i}^{k+1} c(e^1) \cdots c(e^{2k}) , \tag{3.6}$$

where $b^{1\#}, \dots, b^{k\#}$ act on S and generate $\mathcal{F}_k$, we find that $c(e^1), \dots, c(e^n)$ define a representation of $\mathrm{Cl}(V)$ on S. Eqs. (3.5), (3.6) define the *unique*, up to a sign related to space reflection, *irreducible unitary* representation of $\mathrm{Cl}(V)$, which is faithful. Hence

$$\mathrm{Cl}(V) \simeq M\left(2^k, \mathbb{C}\right) . \tag{3.7}$$

A scalar product on $\mathrm{Cl}(V)$ extending the one on V is defined by

$$\langle a , b \rangle := 2^{-k} \mathrm{tr}(a^* b) , \tag{3.8}$$

$a, b \in \mathrm{Cl}(V)$.

The *spin group* $\mathrm{Spin}(V)$ is defined by

$$\mathrm{Spin}(V) := \left\{ a \in \mathrm{Cl}_{\mathbb{R}}^{\mathrm{even}}(V) \,\middle|\, aa^* = a^* a = \mathbb{1}, ac(V)a^* \subseteq c(V) \right\} , \tag{3.9}$$

where $\mathrm{Cl}_{\mathbb{R}}^{\mathrm{even}}(V)$ denotes the real subalgebra of $\mathrm{Cl}(V)$ generated by products of an even number of elements of the form $c(v)$, $v \in V$. We also set $\mathrm{Spin}(n) = \mathrm{Spin}(\mathbb{E}^n)$. The group $\mathrm{Spin}^{\mathbb{C}}(V)$ is defined by

$$\mathrm{Spin}^{\mathbb{C}}(V) := \left\{ e^{\mathrm{i}\alpha} a \,\middle|\, \alpha \in \mathbb{R},\, a \in \mathrm{Spin}(V) \right\} . \tag{3.10}$$

For each $a \in \mathrm{Spin}^{\mathbb{C}}(V)$, we define a linear transformation $\mathrm{Ad}(a)$ of V by

$$c\big(\mathrm{Ad}(a)v\big) := ac(v)a^* , \; v \in V . \tag{3.11}$$

Clearly, this linear transformation preserves the scalar product on V, and we have the short exact sequence

$$1 \longrightarrow \mathrm{U}(1) \longrightarrow \mathrm{Spin}^{\mathbb{C}}(V) \xrightarrow{\mathrm{Ad}} \mathrm{SO}(V) \longrightarrow 1 .$$

The Lie algebra $\mathrm{spin}^{\mathbb{C}}(V)$ of $\mathrm{Spin}^{\mathbb{C}}(V)$ is given by

$$\mathrm{spin}^{\mathbb{C}}(V) = \mathrm{spin}(V) \oplus \mathbb{R} , \tag{3.12}$$

where

$$\mathrm{spin}(V) = \left\{ \xi \in \mathrm{Cl}_{\mathbb{R}}^{\mathrm{even}}(V) \middle|\, \xi + \xi^* = 0 , \, [\xi, c(V)] \subseteq c(V) \right\} . \tag{3.13}$$

One then finds that

$$\mathrm{spin}(V) = \Big\{ \sum_{i,j} x_{ij} c(e^i) c(e^j) \,\Big|\, x_{ij} = -x_{ji} \in \mathbb{R} \Big\} \simeq \mathrm{so}(V)\,. \tag{3.14}$$

Given V, let $\bigwedge^{\bullet}(V \otimes \mathbb{C})$ denote the exterior algebra over $V \otimes \mathbb{C}$. There is a canonical scalar product on $\bigwedge^{\bullet}(V \otimes \mathbb{C})$ extending the one on $V \otimes \mathbb{C} = \bigwedge^1(V \otimes \mathbb{C})$. For $v \in V \otimes \mathbb{C}$, we define operators $a^*(v)$ and $a(v)$ on $\bigwedge^{\bullet}(V \otimes \mathbb{C})$ by setting

$$a^*(v)\, w := v \wedge w\,, \tag{3.15}$$

$$a(v)\, w := \imath(G\bar{v})\, w\,, \tag{3.16}$$

where G is the metric on V defining the scalar product on V, so that Gv is in the dual space of V, and $\imath$ denotes interior multiplication. Then $a(v) = (a^*(v))^*$, and the operators $a^*(v)$, $a(v)$, $v \in V \otimes \mathbb{C}$, are the usual fermionic creation- and annihilation operators satisfying the CAR, with

$$\bigwedge^{\bullet}(V \otimes \mathbb{C}) \simeq \text{fermionic Fock space over } V \otimes \mathbb{C}\,. \tag{3.17}$$

The operators

$$\Gamma(v) := a^*(v) - a(v)\,, \qquad \overline{\Gamma}(v) := \mathrm{i}\big(a^*(v) + a(v)\big)\,, \tag{3.18}$$

then define two anti-commuting unitary representations of $\mathrm{Cl}(V)$ on $\bigwedge^{\bullet}(V \otimes \mathbb{C})$.

Let $\dim V = 2k$ $(p = 0)$ be even. We set

$$\gamma = \mathrm{i}^k \Gamma(e^1) \cdots \Gamma(e^n)\,,$$

which *anti-commutes* with all $\Gamma(v)$, and satisfies $\gamma^2 = \mathbb{1}$. Let $S \simeq \mathbb{C}^{2^k} \simeq \overline{S}$. We then have that

$$\bigwedge^{\bullet}(V \otimes \mathbb{C}) \simeq S \otimes \overline{S}\,,$$

with

$$\Gamma(v) \simeq c(v) \otimes \mathbb{1}\,, \tag{3.19}$$

$$\overline{\Gamma}(v) \simeq \gamma \otimes \bar{c}(v)\,, \tag{3.20}$$

where c and $\bar{c}$ denote the irreducible representations of $\mathrm{Cl}(V)$ on S and $\overline{S}$, respectively.

If $\dim V = 2k + 1$ is odd then

$$\gamma = \mathrm{i}^{k+1} \Gamma(e^1) \cdots \Gamma(e^n)$$

commutes with all $\Gamma(v)$, and satisfies $\gamma^2 = \mathbb{1}$. The operator γ has two eigenvalues, ± 1, both with multiplicity 2^{n-1}. It follows that

$$\bigwedge^{\bullet}(V \otimes \mathbb{C}) \simeq S \otimes \mathbb{C}^2 \otimes \overline{S}\,,$$

and

$$\Gamma(v) = c(v) \otimes \tau_3 \otimes \mathbb{1}\,, \tag{3.21}$$

$$\overline{\Gamma}(v) = \mathbb{1} \otimes \tau_1 \otimes \bar{c}(v)\,. \tag{3.22}$$

3.2. Pauli's wave equation for an 'electron' and for 'positronium' in a general differential-geometric formulation – susy QM

We are ready, now, to formulate *Pauli's wave equation for spinning particles* [19] on a space-time $\mathcal{N} = \mathcal{M} \times \mathbb{R}$, where $\mathcal{M}$ is a general, n-dimensional smooth (compact) spin$^{\mathbb{C}}$ manifold, e.g., $\mathcal{M} = \mathbb{E}^n$, $n = 2, 3$. Let $g = (g_{ij})$ be a *Riemannian metric* on the tangent bundle $\mathrm{T}\mathcal{M}$ of $\mathcal{M}$, and let $G = (g^{ij})$ denote the corresponding inverse metric on the cotangent bundle T^*M. Let $\bigwedge^{\bullet}\mathcal{M}$ be the bundle of differential forms on $\mathcal{M}$, with $\Omega^{\bullet}(\mathcal{M})$ the space of *complexified sections* of $\bigwedge^{\bullet}\mathcal{M}$. This space is equipped with a natural scalar product $\langle \cdot, \cdot \rangle$, determined by g and by the Riemannian volume form. Let $\mathrm{Cl}(\mathcal{M})$ be the Clifford bundle over $\mathcal{M}$; its base space is $\mathcal{M}$ and its fibres are given by $\mathrm{Cl}(\mathrm{T}^*_x\mathcal{M}) \simeq \mathrm{Cl}(\mathbb{E}^n)$, with $n = \dim \mathcal{M}$. Let $\mathcal{A} = C^\infty(\mathcal{M})$ be the algebra of smooth functions on $\mathcal{M}$. The space of sections, $\Gamma(E)$, of a vector bundle E over $\mathcal{M}$ is a finitely generated, projective module for $\mathcal{A}$; E is trivial iff $\Gamma(E)$ is a free $\mathcal{A}$-module. Our standard examples for E are

$$E = \mathrm{T}\mathcal{M},\ \mathrm{T}^*\mathcal{M},\ \bigwedge^{\bullet}\mathcal{M},\ \mathrm{Cl}(\mathcal{M}) .$$

The Clifford bundle over $\mathcal{M}$ has two anti-commuting unitary representations, Γ and $\overline{\Gamma}$, on the module $\Omega^{\bullet}(\mathcal{M})$, which we define as follows: Given a (complex) 1-form $\omega \in \Omega^1(\mathcal{M})$, we introduce *creation-* and *annihilation operators*, $a^*(\omega)$ and $a(\omega)$, on $\Omega^{\bullet}(\mathcal{M})$,

$$a^*(\omega)\sigma := \omega \wedge \sigma , \qquad a(\omega)\sigma := \imath(G\omega)\,\sigma , \tag{3.23}$$

for $\sigma \in \Omega^{\bullet}(\mathcal{M})$. Then (with $a^{\#} = a$ or a^*)

$$\{a^{\#}(\omega_1), a^{\#}(\omega_2)\} = 0 , \qquad \{a(\omega_1), a^*(\omega_2)\} = (\omega_1 , \omega_2) , \tag{3.24}$$

for ω_1, $\omega_2 \in \Omega^1(\mathcal{M})$, where $(\cdot, \cdot)$ is the hermitian structure on $\bigwedge^{\bullet}\mathcal{M}$ determined by G. We define two anti-commuting representations Γ and $\overline{\Gamma}$ of $\mathrm{Cl}(\mathcal{M})$ on $\Omega^{\bullet}(\mathcal{M})$ by setting

$$\Gamma(\omega) := a^*(\omega) - a(\omega) , \qquad \overline{\Gamma}(\omega) := \mathrm{i}\big(a^*(\omega) + a(\omega)\big) . \tag{3.25}$$

If the manifold $\mathcal{M}$ is spin$^{\mathbb{C}}$ (which we have assumed) then

$$\Omega^{\bullet}(\mathcal{M}) = \Gamma(S) \otimes_{\mathcal{A}} \big(\mathbb{C}^2 \otimes\big)\, \Gamma(\overline{S}) , \tag{3.26}$$

where $S \equiv S(\mathcal{M})$ is the *spinor bundle* and $\overline{S}$ the (charge-) conjugate spinor bundle over $\mathcal{M}$. The factor $\mathbb{C}^2$ on the R.S. of (3.26) only appears if $n = \dim \mathcal{M}$ is odd. The modules $\Gamma(S)$ and $\Gamma(\overline{S})$ carry unitary representations c and $\bar{c}$, respectively, of $\mathrm{Cl}(\mathcal{M})$ with

$$\Gamma(\omega) = c(\omega) \otimes \big(\tau_3 \otimes\big)\, \mathbb{1} , \tag{3.27}$$

$$\overline{\Gamma}(\omega) = \gamma \otimes \big(\tau_1 \otimes\big)\, \bar{c}(\omega) , \tag{3.28}$$

with $\gamma = \mathbb{1}$ if n is odd; see Sect. 3.1. (Over a coordinate chart of $\mathcal{M}$, Eqs (3.26)–(3.28) always make sense, by the results of Sect. 3.1. But, *globally*, they *only* make sense if $\mathcal{M}$ is spin$^{\mathbb{C}}$!)

Let ∇ be the *Levi-Civita connection* on $\bigwedge^{\bullet}\mathcal{M}$ (unitary with respect to g and torsion-free). A connection ∇^S on S is called a spin$^{\mathbb{C}}$ connection iff it satisfies the '*Leibniz rule*'

$$\nabla^S_X\big(c(\xi)\psi\big) = c(\nabla_X\xi)\psi + c(\xi)\nabla^S_X\psi\,, \tag{3.29}$$

where X is a vector field on $\mathcal{M}$, ξ a 1-form and ψ a *spinor* in $\Gamma(S)$, i.e., a section of S.

If ∇^S_1 and ∇^S_2 are two hermitian spin$^{\mathbb{C}}$ connections on S then

$$\left(\nabla^S_1 - \nabla^S_2\right)\psi = \mathrm{i}\alpha\otimes\psi\,, \tag{3.30}$$

for $\psi\in\Gamma(S)$, where α is a real, globally defined 1-form. Physically, α is the difference of two *electromagnetic vector potentials*, A_1 and A_2, so-called '*virtual* U(1)-*connections*' on S; (A_i, $i=1,2$, is 'one half times a U(1)-connection' on a line bundle, canonically associated with $S\otimes S$, with magnetic monopoles inside non-contractible 2-spheres in the homology of $\mathcal{M}$).

Given a spin$^{\mathbb{C}}$ connection ∇^S corresponding to a virtual U(1)-connection A, the *Pauli (-Dirac) operator* D_A associated with ∇^S on S is defined by

$$D_A := c\circ\nabla^S\,, \tag{3.31}$$

which is a linear operator on $\Gamma(S)$. Locally, in a coordinate chart of $\mathcal{M}$, with coordinates $x^1,\dots,x^n$,

$$D_A = \sum_{i=1}^{n} c(\mathrm{d}x^i)\,\nabla^S_i\,, \tag{3.32}$$

with

$$\left\{c(\mathrm{d}x^i), c(\mathrm{d}x^j)\right\} = g^{ij}(x)\,.$$

To every ∇^S there corresponds a unique conjugate connection $\overline{\nabla}^S$ on $\overline{S}$, obtained by reversing the electric charge, i.e., $A\to -A$, and we define

$$\overline{D}_{-A} := \bar{c}\circ\overline{\nabla}^S\,, \tag{3.33}$$

an operator acting on $\Gamma(\overline{S})$.

The bundles S and $\overline{S}$ are equipped with a natural hermitian structure. Let dvol_g denote the Riemannian volume form on $\mathcal{M}$. By $\mathscr{H}_\mathrm{e}$ we denote the Hilbert-space completion of $\Gamma(S)$ in the scalar product on $\Gamma(S)$ determined by the hermitian structure of S and dvol_g; $\mathscr{H}_\mathrm{p}$ is defined similarly, with S replaced by $\overline{S}$.

We note, in passing, that the closures of D_A, $\overline{D}_{-A}$ are selfadjoint, elliptic operators densely defined on $\mathscr{H}_\mathrm{e}$, $\mathscr{H}_\mathrm{p}$, respectively.

Thus, $\mathcal{M}$ equipped with a Riemannian metric g, gives rise to what *Alain Connes* [23] calls *spectral triples*

$$(\mathcal{A},\, D_A,\, \mathscr{H}_\mathrm{e})\,,\ (\mathcal{A},\, \overline{D}_{-A},\, \mathscr{H}_\mathrm{p})\,, \tag{3.34}$$

which, in turn, determine $(\mathcal{M},\, g)$ uniquely. In the special case where $\mathcal{M}=\mathbb{E}^3$, these spectral triples are familiar to anyone who knows *Pauli's non-relativistic quantum theory of the spinning electron* and its twin, the positron: $\mathcal{A}$ is the algebra of position measurements; $\mathscr{H}_{\mathrm{e}}$ ($\mathscr{H}_{\mathrm{p}}$) is the Hilbert space of pure state vectors of a single electron (positron); and D_A $\left(\overline{D}_{-A}\right)$ is the 'square-root' of the *Hamiltonian* generating the unitary time evolution of states of an electron (positron) moving in $\mathcal{M}$ and coupled to an external magnetic field $B = \mathrm{d}A$. More precisely, the Hamiltonian is given by

$$H_A = \frac{\hbar^2}{2m} D_A^2 \,, \tag{3.35}$$

where m is the mass of an electron, $\hbar$ is Planck's constant, and the gyromagnetic factor $g = g_{\mathrm{e}} = 2$. (If g_{e} were *different* from 2 then H_A would *not* be the square of D_A; there would then appear an additional *Zeeman term* on the R.S. of (3.35), as Pauli had introduced it in [19]. This term is proportional to $B_{ij}\, c(\mathrm{d}x^i)c(\mathrm{d}x^j)$, in local coordinates, where B is the field strength corresponding to A.) In the presence of an electrostatic potential Φ, the Hamiltonian of Pauli's non-relativistic electron is given by

$$H_{(\Phi,A)} := H_A + \Phi \,, \tag{3.36}$$

and Pauli's version of the time-dependent Schrödinger equation reads

$$\mathrm{i}\hbar \frac{\partial}{\partial t} \psi_t = H_{(\Phi,A)}\, \psi_t \,, \tag{3.37}$$

for $\psi_t \in \mathscr{H}_{\mathrm{e}}$. The corresponding equation for the non-relativistic positron is

$$\mathrm{i}\hbar \frac{\partial}{\partial t} \chi_t = \left(\frac{\hbar^2}{2m} \overline{D}\,{}_{-A}^{\,2} - \Phi \right) \chi_t \,, \tag{3.38}$$

for $\chi_t \in \mathscr{H}_{\mathrm{p}}$.

We observe that when the electrostatic potential Φ vanishes $H_{(0,A)} = H_A$ is the square of a selfadjoint operator (a '*super charge*')

$$Q := \sqrt{\frac{\hbar^2}{2m}}\, D_A \,.$$

Let the dimension of $\mathcal{M}$ be even, and let $\{\varepsilon_1, \ldots, \varepsilon_n\}$ be a local, orthonormal basis of $\Omega^1(\mathcal{M})$; $\left(\{\varepsilon_1, \ldots, \varepsilon_n\}\right.$ is called an 'n-bein'$\left.\right)$. We set

$$\gamma := \mathrm{i}^{\frac{n}{2}}\, c(\varepsilon_1) \cdots c(\varepsilon_n) \,.$$

Since $\mathcal{M}$ is orientable, γ extends to a globally defined involution of $\mathrm{Cl}(\mathcal{M})$ anti-commuting with $c(\omega)$, $\omega \in \Omega^1(\mathcal{M})$, and hence with Q. Then $(\gamma,\, Q,\, \mathscr{H}_{\mathrm{e}})$ furnishes an example of *supersymmetric quantum mechanics*, with $N = 1$ (or (1,0)) super-symmetry. The 'super trace'

$$\mathrm{tr}_{\mathscr{H}_{\mathrm{e}}} \left(\gamma\, \mathrm{e}^{-\beta Q^2} \right) , \quad \beta > 0 \,, \tag{3.39}$$

is easily seen to be independent of β and invariant under small deformations of the metric g and the vector potential A. It computes the *index* of the '*Dirac operator*' D_A, which is a topological invariant of $\mathcal{M}$.

Next, we study the quantum theory of *positronium*, namely of a bound state of an electron and a positron. We define $\mathscr{H}_{\mathrm{e-p}}$ to be the Hilbert space completion of the space $\Omega^{\bullet}(\mathcal{M})$ of differential forms in the scalar product determined by the metric g. Then

$$\mathscr{H}_{\mathrm{e-p}} \simeq \mathscr{H}_{\mathrm{e}} \otimes_{\mathcal{A}} \left(\mathbb{C}^2 \otimes\right) \mathscr{H}_{\mathrm{p}} , \tag{3.40}$$

where the factor $\mathbb{C}^2$ is absent if $\dim \mathcal{M}$ is even. We introduce two anti-commuting Pauli (-Dirac) operators $\mathcal{D}$ and $\overline{\mathcal{D}}$ (densely defined and selfadjoint on $\mathscr{H}_{\mathrm{e-p}}$):

$$\mathcal{D} := \Gamma \circ \nabla , \qquad \overline{\mathcal{D}} := \overline{\Gamma} \circ \nabla , \tag{3.41}$$

where ∇ is the Levi-Civita connection on $\Omega^{\bullet}(\mathcal{M})$, and Γ, $\overline{\Gamma}$ are the two anti-commuting representations of $\mathrm{Cl}(\mathcal{M})$ on $\Omega^{\bullet}(\mathcal{M})$ introduced in (3.23) - (3.25). These operators are easily seen to satisfy

$$\left\{\mathcal{D}, \overline{\mathcal{D}}\right\} = 0 , \qquad \mathcal{D}^2 = \overline{\mathcal{D}}^2 . \tag{3.42}$$

Setting

$$\mathrm{d} := \frac{1}{2}\left(\mathcal{D} - \mathrm{i}\overline{\mathcal{D}}\right) , \qquad \mathrm{d}^* := \frac{1}{2}\left(\mathcal{D} + \mathrm{i}\overline{\mathcal{D}}\right) , \tag{3.43}$$

we find that $\mathrm{d}^2 = \left(\mathrm{d}^*\right)^2 = 0$. In fact, d turns out to be the exterior derivative. The Hamiltonian (for the center-of-mass motion of the 'groundstates' of a bound electron-positron pair, i.e.,) of *positronium* is given by

$$H := \frac{\hbar^2}{2\mu}\mathcal{D}^2 = \frac{\hbar^2}{2\mu}\overline{\mathcal{D}}^2 = \frac{\hbar^2}{2\mu}\left(\mathrm{d}\mathrm{d}^* + \mathrm{d}^*\mathrm{d}\right) , \tag{3.44}$$

where $\mu = 2m$. Note that $\mathcal{D}$, $\overline{\mathcal{D}}$ and H are *independent* of the choice of the vector potential A (and of Φ) which, physically, corresponds to the circumstance that the electric charge of positronium is zero. The data $\left(\mathcal{A}, \mathcal{D}, \overline{\mathcal{D}}, \mathscr{H}_{\mathrm{e-p}}\right)$ are thus well defined even if $\mathcal{M}$ does *not* admit a $\mathrm{spin}^{\mathbb{C}}$ structure. These data, together with (3.44), furnish an example of supersymmetric quantum mechanics with $N = (1,1)$ supersymmetry; the supercharges are the operators $\mathcal{D}$ and $\overline{\mathcal{D}}$. They completely encode the *de Rham-Hodge theory* and the *Riemannian geometry* of $\mathcal{M}$.

One may wonder how additional geometric structure of $\mathcal{M}$ reveals itself in Pauli's quantum theory of a non-relativistic electron, positron or positronium moving in $\mathcal{M}$. Suppose, e.g., that $\mathcal{M}$ is a symplectic manifold equipped with a symplectic 2-form ω. Let Ω denote the anti-symmetric bi-vector field associated with ω. We define three operators on $\mathscr{H}_{\mathrm{e-p}}$

$$L_3 := T - \frac{n}{2} , \quad L_+ := \frac{1}{2}\omega \wedge (\,\cdot\,) , \quad L_- := \frac{1}{2}\imath(\Omega) , \tag{3.45}$$

where $T\lambda = p\,\lambda$, for any p-form $\lambda \in \Omega^{\bullet}(\mathcal{M})$. Then

$$\left[L_3, L_\pm\right] = \pm 2L_\pm , \qquad \left[L_+, L_-\right] = L_3 , \tag{3.46}$$

i.e. $\left\{L_3,\, L_+,\, L_-\right\}$ define a representation of the Lie algebra sl_2 on $\mathscr{H}_{\mathrm{e-p}}$ commuting with the representation of the algebra $\mathcal{A}$ on $\mathscr{H}_{\mathrm{e-p}}$. It is actually a unitary representation, because ${L_3}^* = L_3$ and $(L_\pm)^* = L_\mp$, in the scalar product of $\mathscr{H}_{\mathrm{e-p}}$.

Since ω is closed, we have that $[L_+, \mathrm{d}] = 0$, where d is the exterior derivative. A differential $\tilde{\mathrm{d}}^*$ of degree -1 can be defined by

$$\tilde{\mathrm{d}}^* := [L_-, \mathrm{d}] . \tag{3.47}$$

One finds that $\{\tilde{\mathrm{d}}^*, \mathrm{d}\} = 0$, $(\tilde{\mathrm{d}}^*)^2 = 0$, and $[L_-, \tilde{\mathrm{d}}^*] = 0$. Thus $(\mathrm{d}, \tilde{\mathrm{d}}^*)$ transforms as a doublet under the adjoint action of sl_2.

One can introduce a second sl_2 doublet, $(\tilde{\mathrm{d}}, -\mathrm{d}^*)$, of differentials with the same properties as $(\mathrm{d}, \tilde{\mathrm{d}}^*)$. We are *not* claiming that $\{\mathrm{d}, \tilde{\mathrm{d}}\} = 0$; this equation does *not* hold for general symplectic manifolds. It is natural to ask, however, what is special about the geometry of $\mathcal{M}$ if

$$\{\mathrm{d}, \tilde{\mathrm{d}}\} = 0 . \tag{3.48}$$

It turns out that, in this case, $\mathcal{M}$ is a *Kähler manifold.* Defining

$$\partial := \frac{1}{2}\left(\mathrm{d} - \mathrm{i}\,\tilde{\mathrm{d}}\right), \qquad \bar{\partial} := \frac{1}{2}\left(\mathrm{d} + \mathrm{i}\,\tilde{\mathrm{d}}\right),$$

one finds that

$$\partial^2 = \bar{\partial}^2 = 0, \qquad \{\partial, \bar{\partial}^{\#}\} = 0, \qquad \{\partial, \partial^*\} = \{\bar{\partial}, \bar{\partial}^*\} .$$

The differentials ∂ and $\bar{\partial}$ are the *Dolbeault differentials.* The complex structure J on $\mathcal{M}$ generates a U(1)-symmetry on the differentials:

$$[J, \mathrm{d}] = -\mathrm{i}\,\tilde{\mathrm{d}}, \qquad [J, \tilde{\mathrm{d}}] = \mathrm{i}\,\mathrm{d} .$$

J commutes with the representation of the algebra $\mathcal{A} = C^\infty(\mathcal{M})$ on $\mathscr{H}_{\mathrm{e-p}}$.

The data $\left(\mathcal{A}, \partial, \partial^*, \bar{\partial}, \bar{\partial}^*, \mathscr{H}_{\mathrm{e-p}}\right)$ furnish an example of a supersymmetric quantum theory with $N = (2,2)$ supersymmetry. If the sl_2-symmetry is broken, but the U(1)-symmetry generated by J is preserved then $\mathcal{M}$ may not be symplectic, but it is a *complex-hermitian* manifold.

It is possible to reformulate all special geometries of smooth manifolds in terms of the supersymmetric quantum mechanics of a non-relativistic electron or of positronium by analyzing the adjoint action of symmetries on the Pauli (-Dirac) operators D_A, $\overline{D}_{-A}$, $\mathcal{D}$ and $\overline{\mathcal{D}}$. This mathematical theme is developed in [20]. The upshot of that analysis is that the non-relativistic quantum mechanics of the spinning electron and of positronium encodes the differential geometry and topology of Riemannian manifolds $\mathcal{M}$ ('physical space') in a perfect manner. There is a complete dictionary between the *geometry* of $\mathcal{M}$ and the *supersymmetries* of the quantum theory.

What about the non-relativistic quantum mechanics of particles with 'higher spin'? Let $(\mathcal{M}, g)$ be an n-dimensional, oriented, smooth, Riemannian manifold with Riemannian metric g and volume form dvol_g. Let ρ be a finite-dimensional, unitary representation of Spin(n) on a Hilbert space V_ρ. If ρ is a double-valued representation of SO(n), i.e., $\sigma(\rho) = -1$, then $\mathcal{M}$ must be assumed to be $\mathrm{spin}^{\mathbb{C}}$; for $\sigma(\rho) = 1$, this assumption is not necessary. From the transition functions of the spinor bundle S (or the tangent bundle $T\mathcal{M}$, for $\sigma(\rho) = 1$) and the representation ρ of $\mathrm{Spin}(n)$ we can construct a hermitian vector bundle E_ρ over $\mathcal{M}$ whose fibres are all isomorphic to V_ρ. The hermitian structure on E_ρ and dvol_g determine a scalar product $\langle \cdot, \cdot \rangle_\rho$ on the space of sections $\Gamma(E_\rho)$. The completion of $\Gamma(E_\rho)$ in the norm determined by the scalar product $\langle \cdot, \cdot \rangle_\rho$ is a Hilbert space $\mathscr{H}_\rho$. A $\mathrm{spin}^{\mathbb{C}}$ connection ∇^S on S (or the Levi-Civita connection ∇ on $\bigwedge^{\bullet}\mathcal{M}$ if $\sigma(\rho) = 1$) determines a connection ∇^ρ on E_ρ. (As a physicist, I think about these matters in coordinate charts U of $\mathcal{M}$, with $E_\rho|_U \simeq U \times V_\rho$, use a little representation theory of $\mathrm{Spin}(n)$ and $\mathrm{spin}(n)$, and glue charts together using the transition functions of S, or $T\mathcal{M}$, respectively, in the representation ρ). The connection ∇^ρ, the hermitian structure on E_ρ and dvol_g determine a *Laplace-Beltrami operator* $-\Delta_{g,A}$ densely defined on $\mathscr{H}_\rho$, (e.g., via the Dirichlet form on $\mathscr{H}_\rho$ determined by ∇^ρ).

Pauli's non-relativistic quantum mechanics for a particle moving in physical space $\mathcal{M}$, with an 'intrinsic angular momentum' described by the representation ρ of $\mathrm{Spin}(n)$, is given in terms of the following data: The Hilbert space of pure state-vectors is given by $\mathscr{H}_\rho$. A real 2-form φ on $\mathcal{M}$ determines a section of the subbundle $\mathrm{spin}(\mathcal{M})$ of $\mathrm{Cl}(\mathcal{M})$, whose fibres are all isomorphic to the Lie algebra $\mathrm{spin}(n) \simeq \mathrm{so}(n)$ of $\mathrm{Spin}(n)$; see (3.14). By $\mathrm{d}\rho$ we denote the representation of $\mathrm{spin}(n)$ on V_ρ.

The *Pauli Hamiltonian* is then given by

$$H_A^\rho = -\frac{\hbar^2}{2m}\Delta_{g,A} + \mu_\rho \mathrm{d}\rho(B) + \Phi\,, \tag{3.49}$$

where m is the mass of the particle, μ_ρ its 'magnetic moment', $B \in \Omega^2(\mathcal{M})$ the curvature ('magnetic field') of the virtual U(1)-connection A (the electromagnetic vector potential), and Φ is an external (electrostatic) potential. The second term on the R.S. of (3.49) is the Zeeman term.

Remarks.

(1) Relativistic corrections (spin-orbit interactions) and a variety of further effects can be described in terms of additive contributions to the (U(1)- and) $\mathrm{Spin}(n)$ connection and further Zeeman terms.

(2) In relativistic field theory on four-dimensional space-time, one encounters *acausality phenomena* in the propagation of fields of spin > 1 minimally coupled to external electromagnetic fields ('*Velo-Zwanziger* phenomenon') [25]. This may shed some light on the question why, in Nature, there do not appear to exist any charged elementary particles of spin > 1. See also section 7.1. It should be noted, however, that the Velo-Zwanziger acausality phenomenon disappears in *locally supersymmetric* field theories [26]. (I thank N. Straumann for pointing this out to me.)

Well, I suppose this is all we might want to know about these general matters, right now.

To conclude this general, mathematical section, I want to specialize to the case where $\mathcal{M} = \mathbb{E}^3$, $\mathrm{Spin}(3) = \mathrm{SU}(2)$, which is what we physicists care about most.

3.3. Back to physics: multi-electron systems, Weyl's theorem, the Dirac equation

We first specialize the material of section 3.2 to the case where $\mathcal{M} = \mathbb{E}^3$. Then $S \equiv S(\mathcal{M})$ and $\bigwedge^{\bullet}(\mathcal{M})$ are trivial bundles, and

$$\mathscr{H}_{\mathrm{e}\,/\,\mathrm{p}} \simeq L^2\left(\mathbb{R}^3,\, \mathrm{d}^3x\right) \otimes \mathbb{C}^2\,, \tag{3.50}$$

the space of *square-integrable, two-component spinors* on $\mathbb{R}^3$. Choosing Cartesian coordinates x^1, x^2, x^3 on $\mathbb{E}^3$, the Pauli (-Dirac) operator D_A takes the form

$$D_A = \sum_{j=1}^{3} \sigma_j \left(-\mathrm{i}\frac{\partial}{\partial x^j} + \frac{e}{\hbar c} A_j(x) \right)\,, \tag{3.51}$$

where $\vec{\sigma} = (\sigma_1,\, \sigma_2,\, \sigma_3)$ are the usual Pauli matrices, and $\vec{A}(x) = (A_1(x),\, A_2(x),\, A_3(x))$ is the electromagnetic vector potential in physical units – whence the factor $\frac{e}{\hbar c}$ multiplying $A_j(x)$ in (3.51), where $-e$ is the charge of an electron and c the speed of light. The Pauli Hamiltonian H_A is given by

$$H_A = \frac{\hbar^2}{2m} D_A^2 + \Phi\,, \tag{3.52}$$

where Φ is an external electrostatic potential.

We easily find that

$$\frac{\hbar^2}{2m} D_A^2 = -\frac{\hbar^2}{2m}\Delta_A + \frac{e}{mc}\vec{S}\cdot\vec{B}\,, \tag{3.53}$$

where Δ_A is the covariant Laplacian, $\vec{S} = \frac{\hbar}{2}\vec{\sigma}$ is the spin operator of an electron, and $\vec{B} = \vec{\nabla} \wedge \vec{A}$ is the magnetic field. Thus, for the 'supersymmetric' Hamiltonian H_A, the gyromagnetic factor g_{e} of the electron has the value 2! As long as spin-orbit interactions can be neglected, i.e., in the absence of heavy nuclei, the Hamiltonian H_A in (3.52) describes the dynamics of a slow electron in an external electromagnetic field with good accuracy. Yet, one may wonder how the relativistic effects of spin-orbit interactions and the Thomas precession modify the expression (3.52) for the Pauli Hamiltonian. From (2.14) and (2.17) we find that H_A must then be replaced by

$$H_A^{\mathrm{SO}} = -\frac{\hbar^2}{2m}\Delta_A^2 + \frac{e}{mc}\vec{S}\cdot\left(\vec{B} - \frac{1}{2}\frac{\vec{v}}{c}\wedge\vec{E}\right) + \Phi\,, \tag{3.54}$$

where the (gauge-invariant) *velocity operator* $\vec{v}$ is given by

$$\vec{v} = \frac{\hbar}{m}\left(-\mathrm{i}\vec{\nabla} + \frac{e}{\hbar c}\vec{A}\right)\,, \tag{3.55}$$

and $-\frac{\hbar^2}{2m}\Delta_A = \frac{m}{2}\vec{v}^2$. We introduce a *spin* (SU(2)-) *connection* $w = (w_0, \vec{w})$ on $S(\mathbb{E}^3)$ in terms of its components in the 'natural orthonormal basis' of sections of $S(\mathbb{E}^3)$:

$$w_0(x) = \mathrm{i}\frac{e}{mc\hbar}\vec{B}(x) \cdot \vec{S}\,, \tag{3.56}$$

$$\vec{w}(x) = -\mathrm{i}\frac{e}{2mc\hbar}\vec{E}(x) \wedge \vec{S}\,. \tag{3.57}$$

We then define *covariant derivatives*,

$$D_0 = \frac{1}{c}\frac{\partial}{\partial t} + \frac{\mathrm{i}}{\hbar c}\Phi' + w_0\,, \tag{3.58}$$

where

$$\Phi' = \Phi - \frac{\hbar^2}{2m}\frac{e^2}{8(mc^2)^2}\vec{E}^2\,, \tag{3.59}$$

(D_0 is the *covariant time derivative*), and

$$\vec{D} = \vec{\nabla} + \mathrm{i}\frac{e}{\hbar c}\vec{A} + \vec{w}\,. \tag{3.60}$$

Here $(\Phi', e\vec{A}\,)$ are the components of an *electromagnetic* U(1)-*connection.* Then the Pauli equation,

$$\mathrm{i}\hbar\frac{\partial}{\partial t}\Psi_t = H_A^{\mathrm{SO}}\,\Psi_t, \quad \Psi_t \in \mathscr{H}_{\mathrm{e}}\,,$$

can be rewritten in a *manifestly* $\mathrm{U}(1) \times \mathrm{SU}(2)_{\mathrm{spin}}$ *gauge-invariant* form

$$\mathrm{i}\hbar c D_0\,\Psi_t = -\frac{\hbar^2}{2m}\vec{D}^2\,\Psi_t\,. \tag{3.61}$$

This observation has been made in [27]; (see also the original papers quoted there). When incorporated into the formalism of quantum-mechanical many-body theory the $\mathrm{U}(1) \times \mathrm{SU}(2)_{\mathrm{spin}}$ gauge-invariance of Pauli's theory has very beautiful and important applications to problems in condensed-matter physics, which are discussed in much detail in [27]. Depending on context, the U(1)- and SU(2)-connections introduced above receive further contributions, e.g., from a *divergence-free velocity field* (quantum mechanics in moving coordinates, with applications, e.g., to superconductivity, super-fluidity, a quantum Hall effect for rotating Bose gases [27], nuclear physics,...), from a non-trivial *spin connection* on $S(\mathbb{E}^3)$ with *curvature* and *torsion* describing disclinations and dislocations in a microscopic crystalline background, and/or from the '*Weiss exchange field*' describing a magnetic background. It is most regrettable that we cannot enter into all these applications, here. But the reader will find a detailed exposition of these topics in [27].

Next, we recall the quantum theory of a system of many ($N = 1, 2, 3, \ldots$) Pauli electrons. The Hilbert space of pure state vectors of such a system is given by

$$\mathscr{H}^{(N)} = \mathscr{H}_{\mathrm{e}} \wedge \cdots \wedge \mathscr{H}_{\mathrm{e}} \equiv \mathscr{H}_{\mathrm{e}}^{\wedge N}\,, \tag{3.62}$$

where $\mathscr{H}_{\mathrm{e}}$ is given by (3.50), and $\wedge$ denotes an anti-symmetric tensor product. The anti-symmetric tensor product in (3.62) incorporates the *Pauli exclusion principle*. Let $H^{(1)}$ denote the Pauli Hamiltonian for a single electron, as given in (3.52) or (3.54). In applications to atomic, molecular or condensed matter physics, $\Phi(x)$ is the Coulomb potential of the electron in the field of K nuclei with charges $eZ_1, \ldots, eZ_K$, which we shall usually treat, for simplicity, as *static*, (Born-Oppenheimer approximation); i.e.,

$$\Phi(x) = -\sum_{k=1}^{K} \frac{e^2 Z_k}{4\pi |x - X_k|}, \tag{3.63}$$

where x is the position of the electron, and $X_1, \ldots, X_K$ are the positions of the nuclei. Moreover, $\vec{B}$ is an arbitrary external magnetic field, and $\vec{E}(x) \simeq -\frac{1}{e}\vec{\nabla}\Phi(x)$ is the electric field created by the nuclei (regularized or cut-off, for x near $X_1, \ldots, X_K$).

The Hamiltonian for the N electrons is chosen to be

$$H^{(N)} = \sum_{j=1}^{N} \mathbb{1} \wedge \cdots \wedge H^{(1)} \wedge \cdots \wedge \mathbb{1} + V_{\mathrm{C}}(x_1, \ldots, x_N) + V_{\mathrm{C}}^{\mathrm{nuc}}(X_1, \ldots, X_K), \tag{3.64}$$

where, in the jth term of the sum on the R.S. of (3.64), $H^{(1)}$ stands in the jth place (factor), with $\mathbb{1}$'s in other factors, and

$$V_{\mathrm{C}}(x_1, \ldots, x_N) = \sum_{1 \le i < j \le N} \frac{e^2}{4\pi |x_i - x_j|}, \tag{3.65}$$

$$V_{\mathrm{C}}^{\mathrm{nuc}}(X_1, \ldots, X_K) = \sum_{1 \le k < l \le K} \frac{e^2 Z_k Z_l}{4\pi |X_k - X_l|}. \tag{3.66}$$

Properties of the Hamiltonian $H^{(N)}$ (with $H^{(1)}$ as in (3.52) and Φ as in (3.63)) will be studied in the next section.

We observe that the Hilbert space $\mathscr{H}^{(N)}$ is given by

$$\mathscr{H}^{(N)} = \mathrm{P_a}\left(L^2\left(\mathbb{R}^{3N}, \mathrm{d}^{3N}x\right) \otimes \mathbb{C}^{2^N}\right), \tag{3.67}$$

where $\mathrm{P_a}$ denotes the projection onto the subspace of totally anti-symmetric spinor wave functions. In an obvious sense, $\mathscr{H}^{(N)}$ carries a tensor product representation of two representations, V^{orbit} and V^{spin}, of the permutation group $\mathscr{S}_N$ of N symbols, where

$$V^{\mathrm{orbit}}(\pi) = V_\omega(\pi) \otimes \mathbb{1},$$
$$V^{\mathrm{spin}}(\pi) = \mathbb{1} \otimes V_\sigma(\pi), \quad \pi \in \mathscr{S}_N,$$

in the tensor product decomposition (3.67). The projection $\mathrm{P_a}$ selects the alternating representation (multiplication by $\mathrm{sig}(\pi)$, $\pi \in \mathscr{S}_N$) from $V_\omega \otimes V_\sigma$; only tensor products of subrepresentations, V_ω^i and V_σ^j, of V_ω and V_σ, respectively, are in the range of $\mathrm{P_a}$ for which $V_\omega^i(\pi) = \mathrm{sig}(\pi)\, V_\sigma^j(\pi)$, (i.e., V_ω^i is '*associated*' to V_σ^j).

The spin space $\mathbb{C}^{2^N} \simeq (\mathbb{C}^2)^{\otimes N}$ carries the N-fold tensor product representation, ρ, of the spin $s = \frac{1}{2}$ representation of SU(2). This representation is a direct sum of irreducible representations with spin $s = s_0, s_0 + 1, \ldots, \frac{N}{2}$, where $s_0 = 0$ if N is even and $s_0 = \frac{1}{2}$ if N is odd. It *commutes* with the representation V_σ of $\mathscr{S}_N$ on $(\mathbb{C}^2)^{\otimes N}$.

Hermann Weyl has proven the following

Theorem 3.1.

$$(\mathbb{C}^2)^{\otimes N} \simeq \bigoplus_{(\Delta,\, s)} \mathscr{H}_\Delta \otimes \mathscr{H}_s\,, \tag{3.68}$$

with

$$V_\sigma = \bigoplus_{(\Delta,\, s)} \Delta\big|_{\mathscr{H}_\Delta} \otimes \mathbb{1}\big|_{\mathscr{H}_s} \tag{3.69}$$

$$\rho = \bigoplus_{(\Delta,\, s)} \mathbb{1}\big|_{\mathscr{H}_\Delta} \otimes \rho_s\big|_{\mathscr{H}_s}\,, \tag{3.70}$$

where the Δ's are irreducible representations of the group $\mathscr{S}_N$ labeled by Young diagrams with one *or* two *rows and a total of N boxes, and ρ_s is the irreducible representation of* SU(2) *with spin $s \in \{s_0, s_0 + 1, \ldots, \frac{N}{2}\}$. Moreover, in* (3.68), *every Δ and every s occur only* once, *i.e., a Δ on the R.S. of* (3.68) - (3.70) *paired with a spin s is* uniquely determined *by s, $\Delta = \Delta(s)$, and conversely. (The spin $s = s(\Delta)$ corresponding to a representation Δ is given by half the number of columns in the Young diagram of Δ that consist of a single box.)*

Weyl's theorem is a special case of a general theory of '*dual pairs*' of groups; see [28]. Weyl has shown that the groups $\mathscr{S}_N$ and SU(n), $N = 1, 2, 3, \ldots$, $n = 2, 3 \ldots$ are 'dual pairs'. From our previous discussion we understand that a subrepresentation Δ of V_σ can only be paired with a subrepresentation $\overline{\Delta}$ of V_ω given by

$$\overline{\Delta}(\pi) = \mathrm{sig}(\pi)\Delta(\pi), \quad \pi \in \mathscr{S}_N\,,$$

in order for the tensor product representation $\overline{\Delta} \otimes \Delta$ to 'survive' the projection $\mathrm{P_a}$. This, together with Weyl's theorem, implies that the spin s of an N-electron wave function completely determines its symmetry properties under exchange of electron positions or momenta (the 'race' of the orbital wave function) and under exchange of electron spins (the 'race' of the spin wave function). This explains why in the classification of atomic spectra the permutation groups do not appear; (see section 2). In a system of many electrons moving in a shell of an atom or in a crystalline background, one might expect that, by a conspiracy of electron motion (kinetic energy) and Coulomb repulsion between electrons (potential energy) the energies of those states are particularly low that correspond to totally anti-symmetric orbital wave functions, i.e., $\overline{\Delta}(\pi) = \mathrm{sig}(\pi)$, $\pi \in \mathscr{S}_N$. Then the spin wave functions must be totally symmetric, i.e., Δ must be the trivial representation of $\mathscr{S}_N$. This implies that the spin s of such a state is *maximal*, i.e., $s = \frac{N}{2}$ (for N electrons). The expectation described here is at the core of explanations of *Hund's first rule* and of *ferromagnetism*. While, in many situations, this expectation is quite plausible it is still poorly understood, mathematically.

What is missing? Well, maybe, a few comments on *Dirac's relativistic electron equation*. But I will cut this short, since everybody is familiar with it! A nice approach to the Dirac equation can be extracted from the theory of projective, unitary, irreducible representations of the *Poincaré group* $\mathcal{P}_+^\uparrow$, which is the semi-direct product of the group of proper, orthochronous Lorentz transformations of Minkowski space $\mathbb{M}^4$ and the group of space-time translations. The Poincaré group has *two Casimir operators*,

(i)
$$M^2 = P_0^2 - \vec{P}^2 , \tag{3.71}$$
where $P_0 \equiv H$ (the Hamiltonian) is the generator of time-translations, and $\vec{P}$ (the momentum operator) is the generator of space-translations; and

(ii)
$$W_0^2 - \vec{W}^2 , \tag{3.72}$$
where $(W_0, \vec{W})$ is the *Pauli-Lubanski pseudo vector*; see, e.g., [29].

For purposes of quantum physics, we are only interested in projective, unitary representations of $\mathcal{P}_+^\uparrow$ for which $M^2 \geq 0$ and $W_0^2 - \vec{W}^2$ is finite. In an *irreducible*, projective unitary representation of $\mathcal{P}_+^\uparrow$,

$$\begin{aligned} M^2 &= m^2 \mathbb{1} , \\ W_0^2 - \vec{W}^2 &= -m^2 s(s+1)\, \mathbb{1} , \end{aligned}$$

where $m \geq 0$ is the mass of the representation and (for $m > 0$) s is the *spin* of the representation of the subgroup of space rotations. All projective, unitary, irreducible representations of $\mathcal{P}_+^\uparrow$ corresponding to a given mass $m \geq 0$ and a finite s can be constructed by the method of *induced representations* developed by Wigner and generalized by *George Mackey*. We consider an energy-momentum vector $p = (p_0, \vec{p})$ with $p^2 = p_0^2 - \vec{p}^2 = m^2$. By $\mathrm{H_p}$ we denote the subgroup of all those Lorentz transformations that leave p fixed. For $m > 0$,

$$\mathrm{H_p} \simeq \mathrm{SO}(3) ,$$

while, for $m = 0$,

$$\mathrm{H_p} \simeq \mathrm{E}(2) ,$$

the group of Euclidian motions of the plane. Representations of SO(3) and E(2) then determine representations of $\mathcal{P}_+^\uparrow$. The Hilbert space of pure state vectors of a free, relativistic particle of mass $m \geq 0$ is the representation space of an irreducible unitary representation of the quantum-mechanical Poincaré group with mass $m \geq 0$ and a finite eigenvalue for $W_0^2 - \vec{W}^2$. For an electron or positron, m is positive, and hence $W_0^2 - \vec{W}^2 = -m^2 s(s+1)\, \mathbb{1}$, where s is the spin of the representation of the little group $\mathrm{H_p} \simeq \mathrm{SO}(3)$. For the electron or positron, $s = \frac{1}{2}$!

If we insist that space reflections should be a symmetry of the theory, we must glue together two unitary, irreducible representations of the quantum-mechanical Poincaré group with $m > 0$ and $s = \frac{1}{2}$. Considering that p_0 can be $\geq m$ or $\leq -m$, we find the *Dirac equation* for the relativistic electron hiding in the representation theory of $\mathcal{P}_+^\uparrow$ with mass $m > 0$ and spin $s = \frac{1}{2}$. The second-quantized Dirac theory for free electrons and positrons is obtained by considering anti-symmetric tensor products of the positive-energy representation of $\mathcal{P}_+^\uparrow$ for single electrons and positrons in a rather standard fashion; see, e.g., [29]. All this is so exceedingly well-known that I do not want to enter into details. Similar results can be derived for massless particles ($m = 0$), with spin s replaced by "helicity", λ, with values in (1/2)Z. (I am grateful to my friend R. Stora for education in this matter.)

The results and methods just alluded to, above, can be generalized to Minkowski space-times of arbitrary dimension $d = n + 1 \geq 2$. *Formally*, a *local quantum field theory* of electrons and positrons moving in quite general *Lorentzian* space-time manifolds and coupled to external electromagnetic fields can be written down without difficulty. However, in contrast to the theory of Pauli electrons and positrons moving in a general physical space, the number of electrons and positrons is no longer conserved (electron-positron pair creation processes happen), and one encounters serious analytical problems when one attempts to develop Dirac theory on general Lorentzian space-times and coupled to general electromagnetic fields. These problems are only partially solved, and I do not wish to enter into this matter.

Pauli's non-relativistic theory of the spinning electron, along with a systematic treatment of relativistic corrections, can be recovered by studying the limit of Dirac's theory, as the speed of light c tends to ∞. Relativistic corrections can be found by perturbation theory in c^{-1}. A mathematically careful treatment of such matters can be found in [30].

4. Stability of Non-Relativistic Matter in Arbitrary External Magnetic Fields

In order to get a first idea of the importance of *electron spin* and the *Pauli principle* in the physics of systems of many electrons moving in the Coulomb field of static (light) nuclei and coupled to an arbitrary external magnetic field, I review some fairly recent results on the *stability* of such systems. *The reference* for such results is [31].

Let us consider a system of N electrons and K static nuclei with nuclear charges $eZ_1, \ldots, eZ_k$. with $\sum_{k=1}^{K} Z_k \sim N$. The Hilbert space of the system is the space $\mathscr{H}^{(N)}$ introduced in (3.62), the Hamiltonian is the operator $H^{(N)}$ defined in (3.64), where the one-electron operator $H^{(1)}$ is the Pauli operator of Eq. (3.52), with D_A as in (3.51) and Φ as in (3.63).

Units: The energy unit is $\mathrm{Ry} = 2mc^2\alpha^2$, where $\alpha = \frac{e^2}{\hbar c} \sim \frac{1}{137}$ is Sommerfeld's fine structure constant. The unit of length is half the Bohr radius, i.e., $l = \frac{\hbar^2}{2me^2}$. The magnetic field $\vec{B} = \vec{\nabla} \wedge \vec{A}$ is in units of $\frac{e}{l^2\alpha}$; the magnetic field energy is given by $\varepsilon \int \vec{B}^2 \, \mathrm{d}^3x$, with $\varepsilon = \frac{1}{2\alpha^2}$.

The Pauli operator D_A is given, in our units, by

$$D_A = \vec{\sigma} \cdot \left(-\mathrm{i}\vec{\nabla} + \vec{A} \right) . \tag{4.1}$$

It is convenient to work in the *Coulomb gauge,*

$$\vec{\nabla} \cdot \vec{A} = 0 . \tag{4.2}$$

For a vector field $\vec{X}$ on $\mathbb{R}^3$ or a spinor $\psi \in L^2(\mathbb{R}^3, \mathrm{d}^3x) \otimes \mathbb{C}^2$, we say that $\vec{X} \in L^p$ ($\psi \in L^p$) iff

$$\begin{aligned} \left(\vec{X} \cdot \vec{X}\right)^{1/2} &\in L^p(\mathbb{R}^3, \mathrm{d}^3x) , \\ (\psi, \psi)^{1/2} &\in L^p(\mathbb{R}^3, \mathrm{d}^3x) . \end{aligned}$$

It is shown in [32] that if $\vec{B}$ has *finite field energy,* i.e., $\vec{B} \in L^2$, then there exists a *unique* $\vec{A}$ such that

$$\vec{\nabla} \wedge \vec{A} = \vec{B} , \quad \vec{\nabla} \cdot \vec{A} = 0 , \quad \vec{A} \in L^6 .$$

4.1. Zero-modes of the Pauli operator

Loss and Yau [33] have proven, by a fairly explicit construction, the following important result:

Theorem 4.1. *There exists a single-electron two-component spinor wave function $\psi \in H^1(\mathbb{R}^3)$ (the usual Sobolev space) and a vector potential $\vec{A} \in L^6$, with $\vec{\nabla} \cdot \vec{A} = 0$ and $\vec{B} = \vec{\nabla} \wedge \vec{A} \in L^2$ such that*

$$D_A \psi = 0 , \tag{4.3}$$

i.e, ψ is a zero-mode of the Pauli operator D_A.

An explicit choice of a magnetic field leading to a zero-mode, in the sense of Eq. (4.3) is

$$\vec{B}(x) = \frac{12}{(1+x^2)^3} \Big[(1-x^2)n + 2\,(n \cdot x)\, x + 2n \wedge x \Big] ,$$

where n is a unit vector.

This result, whose proof we omit, has some rather remarkable consequences that we will discuss next. (The proof relies on a three-dimensional analogue of the celebrated Seiberg-Witten equations.)

4.2. Stability and instability of atoms and one-electron molecules

We consider the Pauli Hamiltonian for a one-electron ion in a general external magnetic field $\vec{B}$ of finite field energy:

$$H_A = D_A^2 - \frac{Z}{4\pi\,|x|}\,. \tag{4.4}$$

Let $E_0\,(\vec{B},\,Z)$ denote the *infimum of the spectrum of* H_A. If $\vec{B}$ is a constant external magnetic field, $\vec{B} = (0,\,0,\,B)$, then it is known from work of *Avron, Herbst* and *Simon* quoted in [32] that

$$E_0\,(\vec{B},\,Z) \sim -\mathrm{const}\,(\ln\,B)^2\,.$$

This implies that $E_0\,(\vec{B}_n,\,Z) \longrightarrow -\infty$ even for a sequence of suitably chosen magnetic fields $\vec{B}_n$ of finite, but ever larger field energy. It is then natural to ask whether

$$E_0\,(\vec{B},\,Z) + \varepsilon \int \mathrm{d}^3x\,|\vec{B}\,(x)|^2 \tag{4.5}$$

is bounded below, uniformly in $\vec{B}$, and for what range of values of the nuclear charge.

The answer is worked out in [32]. We define a convenient space, $\mathcal{C}$, of configurations $(\psi,\,\vec{A})$,

$$\mathcal{C} := \left\{ \left(\psi,\,\vec{A}\right)\,\middle|\,\psi \in H^1(\mathbb{R}^3)\,,\,\|\psi\|_2^2 = 1\,,\,\vec{A} \in L^6\,,\vec{\nabla}\cdot\vec{A} = 0\,,\,\vec{\nabla}\wedge\vec{A} \in L^2 \right\} \tag{4.6}$$

and a space $\mathcal{N}$ of ‘zero modes’,

$$\mathcal{N} := \left\{\,\left(\psi,\,\vec{A}\right)\,|\,\left(\psi,\,\vec{A}\right) \in \mathcal{C}\,,\,D_A\psi = 0\,\right\}. \tag{4.7}$$

We then define a critical nuclear charge Z_c by

$$Z_c := \inf_{(\psi,\,\vec{A})\in\mathcal{N}} \left\{\varepsilon\,\|\vec{B}\|_2^2 \,\big/\, \left\langle\psi\,,\,\tfrac{1}{4\pi|x|}\psi\right\rangle\right\}. \tag{4.8}$$

(Note that, by scaling, the analogue of Z_c vanishes in more than three dimensions.)

The following result has been shown in [32].

Theorem 4.2. *Z_c is positive and finite.*

For $Z > Z_c$,

$$\inf_{\vec{B}\in L^2} \left\{E_0\left(\vec{B},\,Z\right) + \varepsilon\,\|\vec{B}\|_2^2\right\} = -\infty\,.$$

For $Z < Z_c$,

$$\inf_{\vec{B}\in L^2} \left\{E_0\left(\vec{B},\,Z\right) + \varepsilon\,\|\vec{B}\|_2^2\right\} > -\infty\,,$$

and the infimum is a minimum reached for some pair $\left(\psi\,,\vec{A}\right) \in \mathcal{C}$.

Furthermore, the infimum on the R.H.S. of (4.8) *is reached on a pair* $\left(\psi, \vec{A}\right) \in \mathcal{N}$.

In [32], Z_c is estimated for the physical value of the fine structure constant and comes out to be $Z_c \sim 17'900$. Thus, a single-electron ion coupled to an arbitrary magnetic field $\vec{B}$ of finite field energy is *stable* (the total energy is bounded from below) if the nuclear charge Z is smaller than Z_c, while it is *unstable* if $Z > Z_c$. This result crucially depends on the fact that electrons have *spin* and a *magnetic moment* with a *gyromagnetic factor* $g_e = 2$, (as long as radiative (QED) corrections are neglected). If $g_e < 2$ then

$$\inf_{\vec{B} \in L^2} E_0 \left(\vec{B}, Z\right) > -\text{const}\, Z^2 > -\infty ,$$

for *all* values of Z, by Kato's '*diamagnetic inequality*', while for $g_e > 2$, ions would *always* be unstable.

In [34], the results summarized in Theorem 4.2 are extended to many-electron atoms and to a system consisting of a single electron moving in the Coulomb field of arbitrarily many static nuclei, (one-electron molecule in the Born-Oppenheimer approximation). For this purpose, one considers the energy functional

$$\mathcal{E}(\Psi, \vec{B}, \underline{X}, \underline{Z}) := \left\langle \Psi , H^{(N)}_{\vec{A}} \Psi \right\rangle + \varepsilon \|\vec{B}\|_2^2 , \tag{4.9}$$

where $\Psi \in \mathscr{H}^{(N)}$, see (3.62), is an N-electron wave function with $\langle \Psi , \Psi \rangle = 1$, and $H^{(N)}_{\vec{A}} \equiv H^{(N)}$ is the N-electron Hamiltonian introduced in (3.64) - (3.66), with $H^{(1)}$ as in (3.52) and (3.63), (see also (4.4), with $\frac{Z}{4\pi|x|}$ replaced by the Coulomb potential (3.63) of many nuclei). There is an obvious extension of the definition (4.6) of the space $\mathcal{C}$ to an N-electron system. We are interested in studying the lowest possible energy

$$E_0 := \inf_{\substack{(\Psi, \vec{A}) \in \mathcal{C} \\ \underline{X} \in \mathbb{R}^{3K}}} \mathcal{E}(\Psi, \vec{B}, \underline{X}, \underline{Z}) . \tag{4.10}$$

It is shown in [34] that, for $K = 1$ (*one* nucleus) and N arbitrary (arbitrarily many electrons), or for K arbitrary and $N = 1$,

$$E_0 > -\infty ,$$

provided $Z_j < \tilde{Z}_c < \infty$, for all $j = 1, \ldots, K$, *and* provided

$$\alpha < \alpha_c , \tag{4.11}$$

with $0.32 < \alpha_c < 6.7$, i.e., *provided the fine structure constant* α *is sufficiently small.* The bound (4.11) comes from studying 1-electron molecules and is '*real*': If $\alpha > \alpha_c$ there are configurations of K identical nuclei with arbitrary $Z < \tilde{Z}_c = \mathcal{O}(\alpha^{-2})$ such that, for some choice of K, $E_0 = -\infty$, for a 1-electron molecule. Again, *the* crucial role in the proofs of these results is played by the electron spin and the fact that $g_e = 2$!

The punchline in this analysis of stability of non-relativistic matter was reached, a little more than ten years ago, in works of *Charles Fefferman* [35] and of *Elliott H. Lieb*, *Michael Loss* and *Jan Philip Solovej* [36] (whose treatment is considerably simpler than Fefferman's, but came a little later)[13]. It is summarized in the next subsection.

4.3. Stability of matter in magnetic fields

Consider the energy functional $\mathcal{E}\,(\Psi, \vec{B}, \underline{X}, \underline{Z})$ introduced in (4.9) – with N electrons moving in the Coulomb field of K static nuclei at positions $X_1, \ldots, X_K$, with nuclear charges $Z_1, \ldots, Z_K$, and coupled to an arbitrary external magnetic field $\vec{B}$ of finite field energy $\varepsilon\,\|\vec{B}\|_2^2$. Let

$$E_0 \equiv E_0(\alpha, \underline{Z}) := \inf_{\substack{(\Psi,\, \vec{A}) \in \mathcal{C} \\ \underline{X} \in \mathbb{R}^{3K}}} \mathcal{E}(\Psi, \vec{B}, \underline{X}, \underline{Z})\,. \tag{4.12}$$

The following result is proven in [36].

Theorem 4.3. *Suppose that $Z_k \le Z < \infty$, for all $k = 1, \ldots, K$, and that*

$$Z\alpha^2 < 0.041 \quad \textit{and} \quad \alpha < 0.06\,. \tag{4.13}$$

Then

$$E_0(\alpha, \underline{Z}) \ge -C\,(N + K)\,, \tag{4.14}$$

for some finite constant C depending on Z and α, but independent *of N and K.*

Remarks. The bound (4.14) expresses *stability of matter* in the sense that the *energy per particle* (electrons and nuclei) has a lower bound ($\ge -\text{const}\, Z^2 \text{Ry}$) *independent* of the number of electrons and nuclei in the system. This is an expression of *thermodynamic stability* of such systems, which is a pillar on which all of condensed-matter physics rests; ('independence' of condensed-matter physics of nuclear form factors and cut-offs imposed on the magnetic field).

For stability of matter, i.e., for the validity of (4.14), it is crucial that electrons are *fermions*, i.e., that they satisfy Pauli's exclusion principle. In Lieb-Thirring type proofs of stability of matter, the Pauli principle enters in the form of generalized Sobolev inequalities (bounding the electron kinetic energy from below by the Thomas-Fermi kinetic energy) *only valid for fermions*; see [31].

We know from the results in the last two subsections that $E_0(\alpha, \underline{Z}) = -\infty$, i.e., the system becomes *unstable*, if either $Z \gg \alpha^{-2}$ or if α is 'large' ($\alpha > 6.7$). It is somewhat tantalizing that *electron spin and the fact that $g_e = 2$ would render systems of many electrons and nuclei* – as they are studied in atomic, molecular and condensed-matter physics – *unstable if $\alpha > 6.7$ and/or if $Z\alpha^2$* is *very 'large'*. This is reminiscent of the possibility that the *Landau pole* in relativistic QED will descend to the non-relativistic regime if α is large enough.

---

[13] All this work came after ground-breaking work of *Dyson* and *Lenard* in the 1960's, and of Lieb and *Thirring*; see [31] and references given there.

Let us see what the source of the potential instability is! It is actually a short-distance or *ultraviolet instability*: If in the definition of $H^{(N)}_{\vec{A}}$, the electromagnetic vector potential $\vec{A}$ in the Coulomb gauge is replaced by a mollified potential

$$\vec{A}_\kappa(x) := \int \mathrm{d}^3 y\, \kappa(x-y)\vec{A}(y)\,,$$

where κ is an arbitrary positive, smooth function, with $\int \kappa = 1$, (i.e., a smooth approximate δ-function) then the bound

$$E_0(\alpha, \underline{Z}) \geq -C\,(N+K)$$

is true for arbitrary α and Z, but the constant C now depends on κ, and if $\alpha > 6.7$ and/or $Z\alpha^2$ is large enough, then $C = C_\kappa \longrightarrow \infty$, as κ approaches a δ-function. In order to arrive at a deeper understanding of these matters, we should *quantize the electromagnetic field*, too.

5. Electrons Interacting with the Quantized Electromagnetic Field; Radiative Corrections to the Gyromagnetic Factor

It is important to ask what becomes of the results in the last section if the electromagnetic field is treated quantum mechanically. One of my strong scientific interests, during the past fifteen years, has been to find mathematically precise answers to this question; see [37–49], and [50] for a review of some of these and other results.

We return to the Hamiltonian (3.64), i.e.

$$H^{(N)} = \sum_{j=1}^{N} \left\{ \left[\vec{\sigma}_j \cdot \left(-\mathrm{i}\vec{\nabla}_j + \vec{A}(x_j)\right)\right]^2 - \sum_{k=1}^{K} \frac{Z_k}{4\pi|x_j - X_k|} \right\} + \sum_{1\leq i<j\leq N} \frac{1}{4\pi|x_i - x_j|} + \sum_{1\leq k<l\leq K} \frac{Z_k Z_l}{4\pi|X_k - X_l|}\,, \tag{5.1}$$

acting on the N-electron Hilbert space

$$\mathscr{H}^{(N)} = \left(L^2\left(\mathbb{R}^3, \mathrm{d}^3x\right) \otimes \mathbb{C}^2\right)^{\wedge N}. \tag{5.2}$$

We are interested in studying the dynamics of such systems when the electromagnetic field is *quantized*, i.e., electrons can emit and absorb photons. We quantize the electromagnetic field in the Coulomb gauge, i.e.,

$$\vec{\nabla} \cdot \vec{A} = 0\,. \tag{5.3}$$

Then

$$\vec{A}(x) = \frac{1}{(2\pi)^{3/2}} \sum_{\lambda=\pm 1} \int \frac{\mathrm{d}^3 k}{\sqrt{2|k|}} \left[\vec{\varepsilon}_\lambda(k) a^*_\lambda(k) \mathrm{e}^{-\mathrm{i}k\cdot x} + \overline{\vec{\varepsilon}_\lambda(k)} a_\lambda(k) \mathrm{e}^{\mathrm{i}k\cdot x}\right], \tag{5.4}$$

where $a^*_\lambda(k)$, $a_\lambda(k)$ are the usual creation and annihilation operators for a photon with wave vector $k \in \mathbb{R}^3$ and helicity $\lambda = \pm$, satisfying the canonical commutation relations (CCR),

$$[a^\#_\lambda(k), a^\#_\mu(l)] = 0\,, \quad [a_\mu(k), a^*_\lambda(l)] = \delta_{\mu\lambda}\delta^{(3)}(k-l)\,, \tag{5.5}$$

and $\vec{\varepsilon}_\lambda(k) \perp k$, $\lambda = \pm$, are two orthonormal polarization vectors. We consider the Fock representation of the commutation relations (5.5) uniquely characterized by the existence of a vacuum state Ω in which none of the field modes is excited, so that

$$a_\lambda(k)\,\Omega = 0\,, \text{ for all } \lambda \text{ and } k\,, \tag{5.6}$$

and $\langle\Omega,\Omega\rangle = 1$. Fock space $\mathscr{F}$ is the Hilbert space completion of the linear space obtained by applying arbitrary polynomials in creation operators smeared out with square-integrable functions to the vacuum Ω. The Hamiltonian of the free electromagnetic field generating the time evolution of vectors in $\mathscr{F}$ is given, in our units, by the operator

$$\begin{aligned} H_\mathrm{f} :=& \frac{1}{2\alpha^2}\int \mathrm{d}^3x \left\{ : \vec{E}(x)^2 : + : \vec{B}(x)^2 : \right\} \\ =& \alpha^{-2} \sum_{\lambda=\pm} \int \mathrm{d}^3k\; a^*_\lambda(k)|k|a_\lambda(k)\,, \end{aligned} \tag{5.7}$$

where

$$\vec{E}(x) = \frac{1}{(2\pi)^{3/2}} \sum_{\lambda=\pm 1} \int \mathrm{d}^3k \sqrt{\frac{|k|}{2}} \left[\mathrm{i}\vec{\varepsilon}_\lambda(k) a^*_\lambda(k) \mathrm{e}^{-\mathrm{i}k\cdot x} - \mathrm{i}\overline{\vec{\varepsilon}_\lambda(k)} a_\lambda(k) \mathrm{e}^{\mathrm{i}k\cdot x} \right],$$

are the transverse components of the electric field, $\vec{B} = \vec{\nabla} \wedge \vec{A}$ is the magnetic field, the double colons indicate standard Wick ordering, and $\alpha^{-2}|k|$ is the energy of a photon with wave vector k (in our units).

The total Hilbert space of electrons and photons is given by

$$\mathscr{H} := \mathscr{H}^{(N)} \otimes \mathscr{F}\,, \tag{5.8}$$

and the Hamiltonian is given by

$$H := H^{(N)} + \mathbb{1} \otimes H_\mathrm{f}\,. \tag{5.9}$$

Alas, this operator is *ill-defined.* To arrive at a mathematically well defined expression for the Hamiltonian (selfadjoint on $\mathscr{H}$ and bounded from below), we must replace the vector potentials $\vec{A}(x_j)$ on the R.S. of (5.1) by ultraviolet regularized potentials $\vec{A}_\Lambda(x_j)$, $j = 1, \ldots, N$, where

$$\vec{A}_\Lambda(x) = \int \mathrm{d}^3y\, \kappa_\Lambda(x-y)\vec{A}(y)\,,$$

and κ_Λ is the Fourier transform of, e.g., a normalized Gaussian

$$\frac{1}{(2\pi\Lambda^2)^{3/2}}\, \mathrm{e}^{-(|k|^2/2\Lambda^2)}\,,$$

where Λ is an ultraviolet cutoff energy that one may choose to be of the order of the rest energy of an electron. Of course one will ultimately be interested in studying the limit, as $\Lambda \longrightarrow \infty$. This limit is only meaningful if the mass and the chemical potential of an electron are *renormalized.* To study the renormalization theory of the model of quantum electrodynamics (QED) considered in this section, we must replace the Pauli Hamiltonians, $\left[\vec{\sigma}_j \cdot \left(- \mathrm{i}\vec{\nabla}_j + \vec{A}(x_j)\right)\right]^2$ on the R.S. of (5.1) by operators

$$\frac{1}{M_\Lambda}\left[\vec{\sigma}_j \cdot \left(-\mathrm{i}\vec{\nabla}_j + \vec{A}_\Lambda(x_j)\right)\right]^2 + \mu_\Lambda\,, \tag{5.10}$$

for $j = 1, \ldots, N$, where M_Λ is the ratio between the '*bare mass*' of an electron and its observed (physical) mass, and μ_Λ is the bare self-energy (or chemical potential) of an electron. The Hamiltonians obtained after the replacement (5.10) are denoted by $H_\Lambda^{(N)} \equiv H_\Lambda^{(N)}(M_\Lambda, \mu_\Lambda)$, see (5.1), and $H_\Lambda \equiv H_\Lambda(M_\Lambda, \mu_\Lambda)$, see (5.9), respectively. A fundamental question in renormalization theory is whether $M_\Lambda > 0$ and μ_Λ can be chosen to depend on the cutoff energy Λ in such a way that the limiting Hamiltonian

$$H_{\mathrm{ren}} = \text{“}\lim_{\Lambda\to\infty} H_\Lambda\text{”} \tag{5.11}$$

exists as a selfadjoint operator on $\mathscr{H}$.

A mathematically rigorous answer to this question remains to be found. (I rather bet it might be 'no'.) However, there are indications of various kinds as to how to choose M_Λ and μ_Λ and plenty of perturbative calculations (perturbation theory in α), which we briefly summarize next.

(1) Since, in our model of QED, the number of electrons and nuclei is conserved – electron-positron pair creation processes are suppressed – there is no vacuum polarization, and hence the fine structure constant α is independent of Λ.

(2) (Non-rigorous) perturbative renormalization group calculations suggest that

$$M_\Lambda \sim \Lambda^{-(8\alpha/3\pi)+\mathcal{O}(\alpha^2)}\,, \tag{5.12}$$

i.e., the bare mass of an electron must approach 0 like a small inverse power of Λ, as $\Lambda \longrightarrow \infty$; or, in other words, the physical mass of an electron consists entirely of radiative corrections[12].

(3) There are some rather crude bounds on the self-energy μ_Λ:

$$c_1\Lambda^{3/2} \leq \mu_\Lambda \leq c_2\Lambda^{12/7}\,,$$

for constants c_1 and c_2 (but derived under the assumption that $M_\Lambda = 1$); see [50] and references given there.

(4) Perturbatively, a finite Lamb shift is found, as $\Lambda \longrightarrow \infty$, which is in rough agreement with experimental data[12]; (an improved version of Bethe's calculation of 1947).

---

[12]In these calculations, the Zeeman terms in $H_\Lambda^{(N)}$ are *neglected.*

(5) The *gyromagnetic factor* g_e *of the electron* is affected by radiative corrections. In low-order perturbation theory in α, it remains *finite*, as $\Lambda \longrightarrow \infty$, and is given by

$$g_e = 2\Big[1 + \frac{8}{3}\frac{\alpha}{2\pi} + \mathcal{O}(\alpha^2)\Big]\,; \tag{5.13}$$

see [50, 51]. This result should be compared to the value for g_e predicted by perturbative *fully relativistic* QED,

$$g_e = 2\Big[1 + \frac{\alpha}{2\pi} + \mathcal{O}(\alpha^2)\Big]\,, \tag{5.14}$$

where the lowest-order correction, $\frac{\alpha}{2\pi}$, was first calculated by *Julian Schwinger*. Experiment favours Schwinger's result! This can be viewed – if one likes – as a high-precision confirmation of, among other things, the *special theory of relativity*.

No matter whether electrons are treated non-relativistically or relativistically, we find that $g_e > 2$! For a single, freely moving electron with Hamiltonian H_A given by (3.52) (with $\Phi = 0$), this results in a breaking of the '*supersymmetry*' (see section 3.2) of the quantum theory, and the effects of 'supersymmetry breaking' offer a handle on *precision measurements* of $g_e - 2$; (see section 6).

The fact that $g_e > 2$ and the results in section 4 apparently imply that QED with non-relativistic matter ultimately only yields a mathematically meaningful description of physical systems if a (large, but finite) *ultraviolet cutoff* is imposed on the interactions between electrons and photons, no matter how small α is chosen. For large values of α ($\alpha > 6.7$), this theory is expected to exhibit cutoff dependence already at atomic and molecular energies.

The need for an ultraviolet cutoff in QED with non-relativistic matter is reminiscent of the problem of the *Landau pole* in *relativistic* QED.

The following results are non-perturbative and mathematically rigorous:

(6) *Stability of Matter:* For an arbitrary number N of electrons and K static nuclei with nuclear charges $Z_k \le Z < \infty$, for all $k = 1, \ldots, K$ and arbitrary $K < \infty$,

$$H_\Lambda \ge -C_{\alpha,\,Z}\, K\Lambda\,, \tag{5.15}$$

for a finite constant $C_{\alpha,\,Z}$ independent of Λ and K. While (5.15) proves stability of matter if an ultraviolet cutoff Λ is imposed on the theory, the linear dependence on Λ on the R.S. of (5.15) is disastrous, physically speaking. It is not understood, at present, whether a lower bound on $H_\Lambda\,(M_\Lambda\,,\mu_\Lambda)$ can be found that is *uniform* in Λ, provided M_Λ and μ_Λ are chosen appropriately!

Present mathematically rigorous efforts towards understanding QED with non-relativistic matter are therefore aimed at an analysis of $H_\Lambda^{(N)}$, for a fixed ultraviolet cutoff Λ ($\sim$ rest energy of an electron), and at tackling the so-called *infrared problem* that is caused by the masslessness of the photons. Here there has been tremendous progress, during the past fifteen years; see e.g. [37–50].

The most remarkable results that have been found, during the last ten years, are, perhaps, the following ones:

We choose an arbitrary, but fixed ultraviolet cutoff Λ.

(7) Atoms have stable ground states; [43–45].

(8) Excited states of atoms are turned into resonances (meta-stable states) whose energies and widths (inverse life times) can be calculated to arbitrary precision by a constructive and convergent algorithm. These energies and life times agree, to leading order in α, with those first calculated by Bethe in order to explain the Lamb shift, [43, 44].

(9) Scattering amplitudes, S_{fi}, for Rayleigh scattering of photons at atoms (below the ionization threshold) have asymptotic expansions of the form

$$S_{fi} = \sum_{n=0}^{N} \sigma_{fi,\,n}(\alpha)\, \alpha^n + o(\alpha^N)\,,$$

where

$$\lim_{\alpha \to 0} \alpha^\delta \sigma_{fi,\,n}(\alpha) = 0\,,$$

for an arbitrarily small $\delta > 0$. It is expected (and can be verified in examples) that

$$\sigma_{fi,\,n}(\alpha) = \sum_{k=0}^{n} \sigma_{fi,\,n,\,k} \left(\ln \frac{1}{\alpha} \right)^k .$$

The powers of $\ln \frac{1}{\alpha}$ come from infrared singularities that render ordinary perturbation theory *infrared-divergent* in large, but finite orders in α; see [48]. Our results yield, among many other insights, a mathematically rigorous justification of Bohr's frequency condition for radiative transitions.

(10) Infrared-finite, constructive, convergent algorithms have been developed to calculate the amplitudes for ionization of atoms by Laser pulses (unpublished work of the author and *Schlein* based on earlier work by *Fring, Kostrykin* and *Schader*) and for *Compton scattering* of photons at a freely moving electron; see *Pizzo* et al. [49].

Most proofs of the results reviewed in this section rely on complex spectral deformation methods, multi-scale perturbation theory and/or operator-theoretic renormalization group methods; see [43, 44, 48] and references given there.

I now leave this thorny territory and sketch how the gyromagnetic factor of the electron can be measured experimentally.

6. Three Methods to Measure g_e

We have already seen in section 2 that atomic spectroscopy in a magnetic field (Zeeman splittings) offers a possibility to measure the gyromagnetic factor g_e of the electron.

Another possibility originating in condensed-matter physics is to exploit the *Einstein–de Haas effect.*

6.1. The Einstein–de Haas effect (see, e.g., [27])

Consider a cylinder of iron magnetized in the direction of its axis and suspended in such a way that it can freely rotate around its axis. Should this cylinder rotate, then it is advisable to treat the quantum theory of the electrons (and nuclei) in the iron in a rotating frame.

Let $\vec{V}(\vec{y}, t)$ be a (divergence-free) vector field on physical space that generates an incompressible flow $\phi_t : \mathbb{E}^3 \to \mathbb{E}^3$ with the property that $\vec{y} = (y^1, y^2, y^3)$, given by

$$\vec{y} = \phi_t^{-1}(\vec{x}), \tag{6.1}$$

are coordinates in the *moving frame* at time t, with $\vec{x} = (x^1, x^2, x^3)$ the Cartesian laboratory coordinates. If $\vec{V}$ generates space rotations around a point $\vec{x}_0$ in space with a fixed angular velocity $\vec{\omega}$ then

$$\vec{V}(\vec{y}, t) = \vec{\omega} \wedge (\vec{y} - \vec{x_0}). \tag{6.2}$$

The quantum theory of electrons in the moving frame is described by a (in general time-dependent) Hamiltonian

$$\begin{aligned} H^{(N)}_{\vec{V}} = \sum_{j=1}^{N} \Big\{ &\frac{m}{2} (\vec{\sigma}_j \cdot \vec{v}_j)^2 + (g_e - 2)\frac{e}{2mc}\frac{\hbar}{2}\vec{\sigma}_j \cdot \vec{B}(\vec{y}_j, t) \\ &- \frac{e}{c}\vec{A}(\vec{y}_j, t) \cdot \vec{V}(\vec{y}_j, t) - \frac{m}{2}\vec{V}(\vec{y}_j, t)^2 \Big\} \\ &+ U_{\text{Coulomb}}\left(\phi_t(\vec{y}_1), \dots, \phi_t(\vec{y}_N), \vec{X}_1, \dots, \vec{X}_K\right), \end{aligned} \tag{6.3}$$

where the velocity operators $\vec{v}_j$ are given by

$$\vec{v}_j = \frac{\hbar}{m}\left(-\mathrm{i}\vec{\nabla}_j + \frac{e}{\hbar c}\vec{A}(\vec{y}_j, t) + \frac{m}{\hbar}\vec{V}(\vec{y}_j, t)\right), \tag{6.4}$$

and U_{Coulomb} is the total Coulomb potential of electrons and nuclei, expressed in laboratory coordinates. The term $-\frac{m}{2}\vec{V}(\vec{y}_j, t)^2$ appearing in (6.3) is the potential of the *centrifugal force* at the position $\vec{y}_j$ of the jth electron in the moving frame. We observe that in (6.3) and (6.4)

$$\frac{e}{c}\vec{A} \quad \text{and} \quad m\vec{V} \tag{6.5}$$

play perfectly analogous rôles, at least if $g_e = 2$. As one will easily guess, $m\vec{V}$ is the vector potential generating the *Coriolis force*, which can be obtained from the

Lorentz force by replacing $\frac{e}{c}\vec{B} = \frac{e}{c}\vec{\nabla} \wedge \vec{A}$ by $m\vec{\nabla} \wedge \vec{V}$. Note that

$$\frac{m}{2}(\vec{\sigma} \cdot \vec{v})^2 + (g_e - 2)\frac{e}{2mc}\vec{S} \cdot \vec{B} \\ = \frac{1}{2m}\left(-i\hbar\vec{\nabla} + \frac{e}{c}\vec{A} + m\vec{V}\right)^2 + \frac{g_e e}{2mc}\vec{S} \cdot \vec{B} + \vec{S} \cdot \vec{\Omega}\,, \quad (6.6)$$

where $\vec{S} = \frac{\hbar}{2}\vec{\sigma}$ is the spin operator of an electron and $\vec{\Omega} = \vec{\nabla} \wedge \vec{V}$ is twice the vorticity of $\vec{V}$.

What we are describing here is the *quantum-mechanical Larmor theorem*: (see, e.g., [27] for details).

Let us now imagine that a magnetized iron cylinder is initially at rest in the laboratory frame. An experimentalist then turns on a constant external magnetic field $\vec{B}$ in the direction opposite to that of the spontaneous magnetization of the cylinder (parallel to its axis), so as to *demagnetize* the cylinder. This causes an *increase* in the free energy of the cylinder, which can be released in the form of mechanical energy. What is this mechanical energy? Well, the cylinder starts to rotate about its axis with an angular velocity $\vec{\omega}$ it chooses so as to cancel the effect of the external magnetic field $\vec{B}$ as best it can. By formula (6.6), the total Zeeman term in the electron Hamiltonian in the rotating frame, *vanishes* if

$$2\vec{\omega} = \vec{\Omega} = -\frac{g_e e}{2mc}\vec{B} \quad (6.7)$$

and the total vector potential affecting orbital motion of the electrons is then given by $\frac{e}{c}\vec{A} + m\vec{V} = \mathcal{O}(g_e - 2) \simeq 0$. The total Coulomb potential U_{Coulomb} is invariant under the transformation $\vec{x}_j \to \vec{y}_j, \vec{X}_j \to \vec{Y}_j$. Thus, in the moving frame, the free energy of the electrons in a cylinder rotating with an angular velocity $\vec{\omega}$ given by (6.7) is approximately the same as the free energy in the laboratory frame before the field $\vec{B}$ was turned on and $\vec{\omega} = 0$. This explains the Einstein–de Haas effect.

By measuring $\vec{B}$ and $\vec{\omega}$, one can determine g_e!

The *Barnett effect* describes the phenomenon that an iron cylinder can be magnetized by setting it into rapid rotation; (see (6.6)).

Other effects based on the same ideas are encountered in cyclotron physics, two-dimensional electron gases exhibiting the quantum Hall effect, molecular and nuclear physics; see [27] and references given there.

6.2. Accelerator measurement of g_e

Consider an electron circulating in an accelerator ring of radius R. It is kept in the ring by a constant external magnetic field $\vec{B}$ perpendicular to the plane of the ring. Its angular velocity $\vec{\omega}_C \parallel \vec{B}$ is found by balancing the centrifugal with the Lorentz force. Thus, its angular velocity is obtained by solving the equation

$$|\vec{\omega}_C| = \frac{e}{\gamma mc} |\vec{B}|, \tag{6.8}$$

where $\gamma = \left(1 - \frac{|\vec{v}|^2}{c^2}\right)^{-1/2}$, $|\vec{v}| = R\,|\vec{\omega}_C|$.

This means that the velocity $\vec{v}$ of the electron precesses around the direction of $\vec{B}$ with an angular frequency $|\vec{\omega}_C|$ given by (6.8). What does its *spin* $\vec{S}$ do? The precession of $\vec{S}$ around $\vec{B}$ is described by the so-called *Bargmann–Michel–Telegdi* (BMT) *equation*. In the special situation considered here, this equation simplifies to

$$\frac{\mathrm{d}\vec{S}}{\mathrm{d}t} = \frac{e}{mc}\vec{S} \wedge \left(\frac{g_\mathrm{e}-2}{2} + \frac{1}{\gamma}\right)\vec{B}, \tag{6.9}$$

see, e.g., [10]. Thus, the precession frequency of the spin is found to be

$$\vec{\omega}_S = \frac{e}{\gamma mc}\vec{B} + \frac{e}{2mc}(g_\mathrm{e}-2)\vec{B}. \tag{6.10}$$

We find that, for $g_\mathrm{e} = 2$, $\vec{\omega}_S = \vec{\omega}_C$; but if $g_\mathrm{e} \neq 2$ the spin- and velocity precession frequencies differ by an amount

$$\frac{e}{2mc}(g_\mathrm{e}-2)\,|\vec{B}|. \tag{6.11}$$

(If $g_\mathrm{e} > 2$ then the spin precesses faster than the velocity.) By measuring the spin polarization of a bunch of electrons, with the property that, *initially*, their spins were parallel to their velocities, after many circulations around the accelerator ring, one can determine $g_\mathrm{e} - 2$ with very high accuracy.

Of course, the formula for the Thomas precession encountered in section 2 can be found as an application of the general BMT equation. How watertight the derivation of the BMT equation is, mathematically, is still a matter of debate [52].

6.3. Single-electron synchrotron measurement of g_e

Consider a single electron in a constant external magnetic field $\vec{B} = (0, 0, B)$ in the z-direction whose motion in the z-direction is quantized by a confining (electrostatic) potential $\Phi(z)$. The time-independent Schrödinger equation for this particle is

$$H^{(1)}\psi = E\psi, \tag{6.12}$$

where $H^{(1)}$ is given by

$$H^{(1)} = \frac{\hbar^2}{2m}\left(-\mathrm{i}\vec{\nabla} + \frac{e}{\hbar c}\vec{A}(\vec{x})\right)^2 + \frac{g_\mathrm{e} e}{2mc} S^{(3)} B + \Phi(z), \tag{6.13}$$

where $\vec{A}(\vec{x}) = \frac{1}{2}(-yB, xB, 0)$, $\vec{x} = (x, y, z)$. Eq. (6.12) can be solved by separating variables:

$$\psi(x, y, z) = \chi(x, y)\,h(z),$$

where χ is a two-component spinor wave function only depending on x and y, and $h(z)$ is a scalar wave function satisfying

$$\left(-\frac{\hbar^2}{2m}\frac{\mathrm{d}^2}{\mathrm{d}z^2} + \Phi(z)\right) h(z) = \mathcal{E}\, h(z)\,, \tag{6.14}$$

with $h \in L^2(\mathbb{R}, \mathrm{d}z)$. Let $\mathcal{E}_0 < \mathcal{E}_1 < \mathcal{E}_2 < \ldots$ be the energy eigenvalues of the eigenvalue problem (6.14). As shown by *Lev Landau*, the energy spectrum of the operator $H^{(1)}$ is then given by the energies

$$E_{n,\,s,\,k} = \hbar\omega_C\left(n + \frac{1}{2}\right) + \frac{g_{\mathrm{e}}}{2}\hbar\omega_C\, s + \mathcal{E}_k\,, \tag{6.15}$$

where $\omega_C = \frac{|eB|}{mc}$, $n = 0, 1, 2 \ldots$, $s = \pm\frac{1}{2}$, $k = 0, 1, 2 \ldots$, and $\mathcal{E}_k$ as in (6.14). All these eigenvalues are infinitely degenerate. Their eigenfunctions corresponding to a degenerate energy level $E_{n,\,s,\,k}$ can be labeled by the eigenvalues of the z-component, L_z, of the orbital angular momentum operator, which are given by

$$\hbar l, \text{ with } l = -n, -n+1, \ldots, 0, 1, 2 \ldots .$$

We observe that if g_{e} were exactly equal to 2 then

$$E_{n,\,-\frac{1}{2},\,k} = E_{n-1,\,+\frac{1}{2},\,k}\,, \tag{6.16}$$

and

$$E_{0,\,-\frac{1}{2},\,k} = \mathcal{E}_k\,.$$

These equations are an expression of the 'supersymmetry' of Pauli's non-relativistic quantum theory of an electron with $g_{\mathrm{e}} = 2$; (see section 3). If $g_{\mathrm{e}} \neq 2$ this supersymmetry is broken, and we have that

$$E_{m-1,\,\frac{1}{2},\,k} - E_{n,\,-\frac{1}{2},\,k} = \hbar\omega_C(m-n) + \frac{g_{\mathrm{e}}-2}{2}\,\hbar\omega_C\,. \tag{6.17}$$

By measuring such energy differences with great precision in very slow radiative transitions, one can determine g_{e} with astounding accuracy. The life times of the excited states can be made long, and hence the energy uncertainties tiny, by using cavities obeying non-resonance conditions. Very beautiful high-precision measurements of g_{e} based on these ideas have recently been performed by *Gerald Gabrielse* and collaborators; see [53].

7. KMS, Spin and Statistics, CPT

In this last section, we study the general connection between the spin of particles and their quantum statistics – particles with half-integer spin are fermions, particles with integer spin are bosons – and the related connection between the spin of fields and their commutation relations within the framework of local relativistic quantum field theory. Our approach to this subject yields, as a byproduct, a proof of the celebrated CPT theorem, namely of the statement that the product of the discrete operations of charge conjugation (C), space reflection (P) and time reversal (T) is an *anti-unitary symmetry* of any local quantum field theory on an even-dimensional space-time. This symmetry maps *states of matter* onto corresponding *states of anti-matter*. Thus the prediction of the existence of the positron by Dirac and Weyl, on the basis of Dirac's hole theory, can be viewed, in hindsight, as a corollary of the locality of quantized Dirac theory and of the general CPT theorem.

I should mention that in a three-dimensional space-time, e.g., in the physics of two-dimensional electron gases exhibiting the quantum Hall effect, or of films, one may encounter (quasi-) particles with *fractional spin* $\notin \frac{1}{2}\mathbb{Z}$ and a type of '*fractional*' *quantum statistics* described by representations of the *braid groups*, or braid groupoids (originally introduced in mathematics by *Emil Artin*). Moreover, in two- and three-dimensional local quantum field theories, there are fields of fractional spin whose commutation relations give rise to representations of the braid groups or groupoids. It is conceivable that this exotic type of quantum statistics is relevant in the context of the fractional quantum Hall effect, and there are people who hope to exploit it for the purpose of (topological) *quantum computing.*[14]

It may be appropriate to make some sketchy remarks on the history of the discoveries of the connection between spin and statistics, of the CPT theorem and of braid statistics.

The general connection between spin and statistics for free fields was discovered, on the basis of earlier work by Heisenberg and Pauli and by Pauli and Weisskopf, by *Markus Fierz* in 1939, [54]. His result was later rederived more elegantly by Pauli. In axiomatic field theory, a general result was found by Lüders and Zumino; see [55, 56]. A much more general analysis of the statistics of superselection sectors, based on the algebraic formulation of local quantum field theory, was carried out by *Doplicher*, *Haag* and *Roberts*; see [58,59]. They showed that general para-Bose or para-Fermi statistics can always be converted into ordinary Bose or Fermi statistics by introducing 'internal degrees of freedom' on which a compact topological group of internal symmetries acts, and they rederived the general connection between spin and statistics. All these results only hold in space-times of dimension ≥ 4.

The CPT theorem, i.e., the statement that the product of C, P and T is an anti-unitary symmetry of any local, relativistic quantum field theory, was first derived in [60] and then, in its general form, by *Res Jost* in [61]; see also [55, 56]. Based on Jost's analysis and on the KMS condition [62] characterizing thermal equilibrium states, it was rederived in a general setting by *Bisognano* and *Wichmann* [63], who established a connection with Tomita-Takesaki theory [64].

We will see that the general connection between spin and statistics and the CPT theorem are consequences of the fact that the vacuum state of a local relativistic quantum field theory is a *KMS* (equilibrium) *state* for all one-parameter subgroups of the Poincaré group consisting of *Lorentz boosts* in a two-dimensional

[14]An idea probably first suggested by myself.

plane containing a time-like direction. This observation has been made in [63]. Incidentally, it is at the core of the theory of the *Unruh effect.*

Exotic commutation relations between fields carrying 'fractional charges' in local relativistic quantum field theories with soliton sectors in *two* space-time dimensions first appeared in work of *R. Streater* and *I. Wilde* [65] and of the author [66], in the early seventies. (They gave rise to certain abelian representations of the braid groups.) In 1977, *M. Leinaas* and *J. Myrheim* [67] discovered the first example of a system of quantum particles moving in the plane and exhibiting braid (or 'fractional') statistics: Charged point particles carrying magnetic vorticity. The braid statistics of such particles is a consequence of the Aharonov-Bohm effect. Their analysis was generalized in [68] and [69]. Within the context of abelian gauge (Higgs) theories in three dimensions, particles with fractional spin and braid statistics were analyzed in [70]. The general theory of (abelian and *non-abelian*) braid statistics was initiated by the author in [71] and completed in [72, 73], and references given there. A general connection between fractional spin and braid statistics was established in [73], and it was shown that, in local theories in three-dimensional space-time, ordinary Bose or Fermi statistics implies that all spins are integer of half-integer, and that braid statistics implies the breaking of parity (P) and time reversal (T).

7.1. SSC, KMS and CPT

I will now first recall the *connection between spin and statistics* (SSC) in the general framework of *local relativistic quantum field theory* (RQFT), as formalized in the so-called *(Gårding-) Wightman axioms* [55, 56]; (see also [57]). As a corollary, I will then show that the vacuum state of an arbitrary local RQFT is a *KMS* (equilibrium) *state* [62] for any one-parameter group of Lorentz boosts at inverse temperature $\beta = 2\pi$, [63]. The *CPT theorem* and *SSC* turn out to be consequences of the KMS condition.

I will follow methods first introduced in [61,63], and my presentation is similar to that in [74], where various mathematical details can be found.

We consider a local RQFT on Minkowski space $\mathbb{M}^d$, ($d = n + 1$), at zero temperature satisfying the Wightman axioms [55, 56]. Let $\mathscr{H}$ denote the Hilbert space of pure state vectors of the theory and $\Omega \in \mathscr{H}$ the *vacuum vector*. The space $\mathscr{H}$ carries a projective, unitary representation, U, of $\mathcal{P}_+^\uparrow$. We first consider RQFT's with fields localizable in points and transforming covariantly under the adjoint action of U; a more general framework is considered in the next subsection, (see [76] for a general analysis of the localization properties of fields). Let $\Psi_1, \dots, \Psi_N$ be the fields of the theory. Smearing out these fields with test functions in the Schwartz space over $\mathbb{M}^d$, one obtains operators densely defined on $\mathscr{H}$. In fact, $\mathscr{H}$ turns out to be the norm-closure of the linear space obtained by applying arbitrary polynomials in $\Psi_1, \dots, \Psi_N$ (smeared out with Schwartz space test functions) to the vacuum Ω. Let $\Pi \subset \mathbb{M}^d$ be a two-dimensional plane containing a time-like direction. Without loss of generality, we can choose coordinates $x^0, x^1, \dots, x^{d-1}$

in $\mathbb{M}^d$ such that Π is the (x^0, x^1)-coordinate plane. We consider the one-parameter subgroup of Lorentz boosts given by

$$\begin{aligned} x^0_\theta &= \cosh(\theta)x^0 + \sinh(\theta)x^1 \,, \\ x^1_\theta &= \sinh(\theta)x^0 + \cosh(\theta)x^1 \,, \\ x^j_\theta &= x^j \,, \quad \text{for } j \geq 2 \,, \end{aligned} \tag{7.1}$$

with $\theta \in \mathbb{R}$ the rapidity of the boost. Let $M_\Pi = M_\Pi^*$ denote the generator of the boosts (7.1) in the projective, unitary representation U of $\mathcal{P}_+^\uparrow$ on $\mathscr{H}$. To each field Ψ_j of the theory, there is associated a finite-dimensional, irreducible projective representation S_j of the group $\mathrm{L}_+^\uparrow$ of proper, orthochronous Lorentz transformations of $\mathbb{M}^d$ such that

$$\mathrm{e}^{\mathrm{i}\theta M_\Pi}\, \Psi_j(x^0, x^1, \vec{x})\, \mathrm{e}^{-\mathrm{i}\theta M_\Pi} = S_j^{-1}(\theta)\, \Psi_j(x^0_\theta, x^1_\theta, \vec{x}) \,, \tag{7.2}$$

with $\vec{x} = (x^2, \ldots, x^{d-1})$, or, in components,

$$\mathrm{e}^{\mathrm{i}\theta M_\Pi}\, \Psi_j^A(x^0, x^1, \vec{x})\, \mathrm{e}^{-\mathrm{i}\theta M_\Pi} = \sum_B S_j^{-1}(\theta)^A{}_B\, \Psi_j^B(x^0_\theta, x^1_\theta, \vec{x}) \,, \tag{7.3}$$

where Ψ_j^A is the Ath component of Ψ_j.

A theorem due to *Bargmann, Hall* and *Wightman* [55, 56] guarantees that, for an RQFT satisfying the Wightman axioms, the *Wick rotation* from real times to purely *imaginary times* $ct = \mathrm{i}\tau$, $\tau \in \mathbb{R}$, is *always possible.* The vacuum vector Ω turns out to be in the domain of all the operators $\prod_{k=1}^n \hat{\Psi}_{j_k}(x_k)$, where $x_k = (\tau_k, x^1_k, \vec{x}) \in \mathbb{E}^d$ (d-dim. Euclidean space),

$$\hat{\Psi}_j(\tau, x^1, \vec{x}) := \Psi_j(\mathrm{i}\tau, x^1, \vec{x}) = \mathrm{e}^{-\tau H}\, \Psi_j(0, x^1, \vec{x})\, \mathrm{e}^{\tau H} \,, \tag{7.4}$$

with $H \geq 0$ the *Hamiltonian* of the theory, provided that

$$0 < \tau_1 < \tau_2 < \ldots < \tau_n \,; \tag{7.5}$$

see [55, 77]. The Euclidian Green- or *Schwinger functions* are then defined by

$$S^{(n)}(j_1, x_1, \ldots, j_n, x_n) := \left\langle \Omega, \hat{\Psi}_{j_1}(x_1) \cdots \hat{\Psi}_{j_n}(x_n)\, \Omega \right\rangle . \tag{7.6}$$

By Bargmann-Hall-Wightman, the Schwinger functions $S^{(n)}$ are defined on all of

$$\mathbb{E}^{dn}_{\neq} := \left\{ (x_1, \ldots, x_n) \,\middle|\, x_j \in \mathbb{E}^d,\ j = 1, \ldots, n,\ x_i \neq x_j,\ \text{for } i \neq j \right\} . \tag{7.7}$$

It is convenient to introduce polar coordinates, $(\alpha, r, \vec{x})$, with $r > 0$, $\alpha \in [0, 2\pi)$, in the (τ, x^1)-plane by setting

$$\tau = r \sin\alpha,\ x^1 = r\cos\alpha,\ \vec{x} = (x^2, \ldots, x^{d-1}) \,; \tag{7.8}$$

(the angle α is an imaginary rapidity).

Let $\mathscr{S}_+$ denote the Schwartz space of test functions $f(r, \vec{x})$ with support in $\mathbb{R}_+ \times \mathbb{R}^{d-2}$. We define functions $\Phi^{(n)}$ of n angles as follows:

$$\Phi^{(n)}(j_1, f_1, \alpha_1, \ldots, j_n, f_n, \alpha_n) :=$$
$$\int S^{(n)}(j_1, \alpha_1, r_1, \vec{x}_1, \ldots, j_n, \alpha_n, r_n, \vec{x}_n) \prod_{k=1}^{n} f_k(r_k, \vec{x}_k)\, \mathrm{d}r_k\, \mathrm{d}^{d-2}x_k\,. \quad (7.9)$$

As shown in [74] (see also [77]), using Bargmann-Hall-Wightman (see (7.6), (7.7)) – among other things – these functions are given by

$$\Phi^{(n)}(j_1, f_1, \alpha_1, \ldots, j_n, f_n, \alpha_n) = \left\langle \Omega, \hat{\Psi}_{j_1}(f_j, \alpha_1) \cdots \hat{\Psi}_{j_n}(f_n, \alpha_n)\, \Omega \right\rangle, \quad (7.10)$$

provided $\alpha_1 < \alpha_2 < \ldots < \alpha_n$, with $\alpha_n - \alpha_1 < 2\pi$. On the R.S. of (7.10),

$$\hat{\Psi}_j(f, \alpha + \beta) = \mathrm{e}^{-\alpha M_\Pi}\, R_j(\alpha)\, \hat{\Psi}_j(f, \beta)\, \mathrm{e}^{\alpha M_\Pi}\,, \quad (7.11)$$

for arbitrary angles $\alpha > 0$, $\beta \geq 0$ with $\alpha + \beta < \pi$, where

$$R_j(\alpha) := S_j(\mathrm{i}\alpha) \quad (7.12)$$

is the finite-dimensional, irreducible representation of Spin(d) obtained from S_j by analytic continuation in the rapidity. Formally, (7.10) and (7.11) follow from (7.2), (7.3) and (7.6); (the details required by mathematical rigor are a little complicated; but see [63, 74]). We note that the vacuum Ω is invariant under Poincaré transformations; in particular

$$\mathrm{e}^{\mathrm{i}\theta M_\Pi}\, \Omega = \Omega\,, \quad \text{for all } \theta \in \mathbb{C}\,. \quad (7.13)$$

We also note that two points $(\alpha_1, r_1, \vec{x}_1)$ and $(\alpha_2, r_2, \vec{x}_2)$ in $\mathbb{E}^d$ are *space-like separated* whenever $\alpha_1 \neq \alpha_2$. Thus, the local commutation relations of fields at space-like separated points [55, 56, 77] imply that, for $\alpha_k \neq \alpha_{k+1}$,

$$\Phi^{(n)}(\ldots, j_k, f_k, \alpha_k, j_{k+1}, f_{k+1}, \alpha_{k+1}, \ldots)$$
$$= \exp(\mathrm{i}2\pi\theta_{j_k\, j_{k+1}})\, \Phi^{(n)}(\ldots, j_{k+1}, f_{k+1}, \alpha_{k+1}, j_k, f_k, \alpha_k, \ldots)\,, \quad (7.14)$$

for arbitrary $1 \leq k < n$, where, for $d \geq 4$,

$$\theta_{j\,j'} = 0 \bmod \mathbb{Z} \text{ if } \Psi_j \text{ or } \Psi_{j'} \text{ is a } \textit{Bose field}\,, \quad (7.15)$$

$$\theta_{j\,j'} = \frac{1}{2} \bmod \mathbb{Z} \text{ if } \Psi_j \textit{ and } \Psi_{j'} \text{ are } \textit{Fermi fields}\,. \quad (7.16)$$

For details see [77] and [78]. In two space-time dimensions, the statistics of fields localizable in points can be more complicated; see subsection 7.2, and [71–73]. In particular, the phases $\theta_{j\,j'}$ can be arbitrary real numbers, and this is related to the fact that $\mathrm{Spin}(2) = \widetilde{\mathrm{SO}(2)} = \mathbb{R}$, which implies that the *spin* (parity) s_j of a field Ψ_j can be an arbitrary real number. The *spin* (parity) s_j of a field Ψ_j is defined as follows: Since R_j is a finite-dimensional, irreducible representation of Spin(d),

$$R_j(2\pi) = \mathrm{e}^{\mathrm{i}2\pi s_j}\, \mathbb{1}\,, \quad (7.17)$$

where $s_j = 0, \frac{1}{2} \bmod \mathbb{Z}$, for $d \geq 3$, while $s_j \in [0, 1) \bmod \mathbb{Z}$, for $d = 2$.

Given a field index j, we define the 'adjoint' index $\overline{j}$ through the equation

$$\left(\Psi_{j}^{B}(g)\right)^{*}=\Psi_{\overline{j}}^{B}(\overline{g}),\ g\in\mathscr{S}(\mathbb{M}^{d}),\tag{7.18}$$

where A^{*} is the adjoint of the operator A on $\mathscr{H}$ in the scalar product of $\mathscr{H}$.

We are now prepared to prove the general *spin-statistics-connection* (SSC) for fields of a local RQFT localizable in space-time points. We first note that, by (7.11) and (7.18),

$$\begin{aligned}\hat{\Psi}_{j}(f,\alpha)^{*}&=\left(\mathrm{e}^{-\alpha M_{\Pi}}R_{j}(\alpha)\,\hat{\Psi}_{j}(f,0)\,\mathrm{e}^{\alpha M_{\Pi}}\right)^{*}\\&=\mathrm{e}^{\alpha M_{\Pi}}R_{j}(\alpha)^{*}\,\hat{\Psi}_{\overline{j}}(\overline{f},0)\,\mathrm{e}^{-\alpha M_{\Pi}}\\&=R_{j}(\alpha)^{*}R_{\overline{j}}^{-1}(-\alpha)\,\hat{\Psi}_{\overline{j}}(\overline{f},-\alpha)\\&\overset{!}{=}\hat{\Psi}_{\overline{j}}(\overline{f},-\alpha),\end{aligned}$$

by (7.2), (7.3) and (7.18). Thus

$$R_{j}(\alpha)^{*}=R_{\overline{j}}(-\alpha).\tag{7.19}$$

Furthermore, by (7.3), (7.11) and (7.13),

$$\begin{aligned}&\Phi^{(n)}(j_{1},f_{1},\alpha_{1}+\alpha,\ldots,j_{n},f_{n},\alpha_{n}+\alpha)\\&\quad=\left\langle\Omega,\prod_{k=1}^{n}\hat{\Psi}_{j_{k}}(f_{k},\alpha_{k}+\alpha)\,\Omega\right\rangle\\&\quad=\left\langle\Omega,\prod_{k=1}^{n}\left(\mathrm{e}^{-\alpha M_{\Pi}}R_{j_{k}}(\alpha)\hat{\Psi}_{j_{k}}(f_{k},\alpha_{k})\,\mathrm{e}^{\alpha M_{\Pi}}\right)\Omega\right\rangle\\&\quad=R_{j_{1}}(\alpha)\otimes\cdots\otimes R_{j_{n}}(\alpha)\,\Phi^{(n)}(j_{1},f_{1},\alpha_{1},\ldots,j_{n},f_{n},\alpha_{n}),\end{aligned}\tag{7.20}$$

which expresses the rotation covariance of the functions $\Phi^{(n)}$, (a consequence of the *Poincaré covariance* of the fields Ψ_{j} and the *Poincaré invariance* of the vacuum Ω). Thus, using the positivity of the scalar product $\langle\cdot,\cdot\rangle$ on $\mathscr{H}$, we find that, for $0<\alpha<\pi$,

$$\begin{aligned}0&<\left\langle\mathrm{e}^{-\alpha M_{\Pi}}\,\hat{\Psi}_{j}(f,0)\,\Omega,\mathrm{e}^{-\alpha M_{\Pi}}\,\hat{\Psi}_{j}(f,0)\,\Omega\right\rangle\\&\overset{(7.10),(7.11)}{=}R_{\overline{j}}^{-1}(-\alpha)\otimes R_{j}^{-1}(\alpha)\,\Phi^{(2)}(\overline{j},\overline{f},-\alpha,j,f,\alpha)\\&\overset{(7.14),(7.19)}{=}R_{j}^{-1}(\alpha)\otimes R_{\overline{j}}^{-1}(-\alpha)\,\mathrm{e}^{\mathrm{i}2\pi\theta_{\overline{j}j}}\,\Phi^{(2)}(j,f,\alpha,\overline{j},\overline{f},-\alpha)\\&=R_{j}^{-1}(\alpha)\otimes R_{\overline{j}}^{-1}(-\alpha)\,\mathrm{e}^{\mathrm{i}2\pi\theta_{\overline{j}j}}\,\Phi^{(2)}(j,f,\alpha,\overline{j},\overline{f},2\pi-\alpha)\\&\overset{(7.20)}{=}R_{j}^{-1}(\alpha-\pi)\otimes R_{\overline{j}}^{-1}(-\alpha-\pi)\,\mathrm{e}^{\mathrm{i}2\pi\theta_{\overline{j},j}}\,\Phi^{(2)}(j,f,\alpha-\pi,\overline{j},\overline{f},\pi-\alpha)\\&\overset{(7.17)}{=}\mathrm{e}^{\mathrm{i}2\pi\theta_{\overline{j}j}}\,\mathrm{e}^{\mathrm{i}2\pi s_{\overline{j}}}\,R_{j}^{-1}(\alpha-\pi)\otimes R_{\overline{j}}^{-1}(\pi-\alpha)\,\Phi^{(2)}(j,f,\alpha-\pi,\overline{j},\overline{f},\pi-\alpha)\\&\overset{(7.11)}{=}\mathrm{e}^{\mathrm{i}2\pi\theta_{\overline{j}j}}\,\mathrm{e}^{\mathrm{i}2\pi s_{\overline{j}}}\left\langle\mathrm{e}^{(\alpha-\pi)M_{\Pi}}\,\hat{\Psi}_{\overline{j}}(\overline{f},0)\,\Omega,\mathrm{e}^{(\alpha-\pi)M_{\Pi}}\,\hat{\Psi}_{\overline{j}}(\overline{f},0)\,\Omega\right\rangle.\end{aligned}\tag{7.21}$$

Note that the L.S. and the scalar product (3rd factor) on the very R.S. of (7.21) are well defined and *strictly positive*, for $0 < \alpha < \pi$. It then follows that

$$s_j = -s_{\bar{j}} = \theta_{\bar{j}\,j} \mod \mathbb{Z}\,, \tag{7.22}$$

which is the usual connection between spin and statistics:

$$\begin{aligned} s_j \textit{ half-integer} &\longleftrightarrow \Psi_j \text{ a } \textit{Fermi field}\,, \\ s_j \textit{ integer} &\longleftrightarrow \Psi_j \text{ a } \textit{Bose field}\,, \end{aligned} \tag{7.23}$$

and, for $d = 2$,

$$s_j \textit{ fractional} \longleftrightarrow \Psi_j \text{ a field with } \textit{fractional} \text{ (braid) } \textit{statistics.}$$

Next, we show that our results imply that the vacuum Ω is a *KMS state* at inverse temperature $\beta = 2\pi$ for the one-parameter group of Lorentz boosts in the plane Π.

We consider the Schwinger function

$$\Phi^{(n)}(j_1, f_1, \alpha_1, \ldots, j_n, f_n, \alpha_n) = \Big\langle \Omega\,, \prod_{k=1}^{n} \hat{\Psi}_{j_k}(f_k, \alpha_k)\,\Omega \Big\rangle\,, \tag{7.24}$$

for $\alpha_1 < \cdots < \alpha_n$, with $\alpha_n - \alpha_1 < 2\pi$. For simplicity, we assume that $d \geq 3$, so that all spins are half-integer or integer and, by (7.22), only Fermi- or Bose statistics is possible. Then $\Phi^{(n)}(j_1, f_1, \alpha_1, \ldots, j_n, f_n, \alpha_n)$ vanishes, unless an *even* number of the fields $\Psi_{j_1}, \ldots, \Psi_{j_n}$ are Fermi fields. For every $1 \leq m < n$, we define the phase

$$\varphi_m = \sum_{\substack{k=1,\ldots,m \\ l=m+1,\ldots,n}} \theta_{j_k\,j_l}\,, \tag{7.25}$$

with $\theta_{j_k\,j_l}$ as in (7.14).

Using Eqs. (7.15) and (7.16) and the fact that the total number of Fermi fields among $\Psi_{j_1}, \ldots, \Psi_{j_n}$ is even, one easily deduces from the spin statistics connection (7.23) that

$$\varphi_m = \sum_{k=1}^{m} s_{j_k} \mod \mathbb{Z}\,. \tag{7.26}$$

Next, by repeated use of (7.14), we find that

$$\begin{aligned} &\Phi^{(n)}(j_1, f_1, \alpha_1, \ldots, j_n, f_n, \alpha_n) \\ &= e^{i2\pi\varphi_m}\,\Phi^{(n)}(j_{m+1}, f_{m+1}, \alpha_{m+1}, \ldots, j_n, f_n, \alpha_n, j_1, f_1, \alpha_1, \ldots, j_m, f_m, \alpha_m) \\ &\overset{(7.26)}{=} \exp\Big(i2\pi \sum_{k=1}^{m} s_{j_k}\Big)\,\Phi^{(n)}(j_{m+1}, f_{m+1}, \alpha_{m+1}, \ldots, j_1, f_1, \alpha_1, \ldots) \\ &\overset{(7.17)}{=} \mathbb{1} \otimes \cdots \otimes \mathbb{1} \otimes R_{j_1}(2\pi) \otimes \cdots \otimes R_{j_m}(2\pi) \\ &\qquad \cdot \Phi^{(n)}(j_{m+1}, f_{m+1}, \alpha_{m+1}, \ldots, j_1, f_1, \alpha_1 + 2\pi, \ldots, j_m, f_m, \alpha_m + 2\pi)\,. \end{aligned} \tag{7.27}$$

Note that $\alpha_{m+1} < \ldots < \alpha_n < \alpha_1 + 2\pi < \ldots < \alpha_{2m} + 2\pi$, with $\alpha_m + 2\pi - \alpha_{m+1} < 2\pi$) ($\Leftrightarrow \alpha_m < \alpha_{m+1}$). Thus, by (7.24) (applied to the L.S. and the R.S. of (7.27)), we arrive at the identity

$$\Big\langle \Omega\,,\ \prod_{k=1}^{m} \hat{\Psi}_{j_k}(f_k,\, \alpha_k) \prod_{l=m+1}^{n} \hat{\Psi}_{j_l}(f_l,\, \alpha_l)\,\Omega \Big\rangle$$
$$= \Big\langle \Omega\,,\ \prod_{l=m+1}^{n} \hat{\Psi}_{j_l}(f_l,\, \alpha_l) \prod_{k=1}^{m} \Big(\mathrm{e}^{-2\pi M_\Pi}\, \hat{\Psi}_{j_k}(f_k,\, \alpha_k)\, \mathrm{e}^{2\pi M_\Pi} \Big)\,\Omega \Big\rangle\,, \tag{7.28}$$

which is the celebrated *KMS condition.*

Defining

$$\omega(A) := \langle \Omega, A\,\Omega \rangle\,, \tag{7.29}$$

and

$$\tau_\theta(A) := \mathrm{e}^{\mathrm{i}\theta M_\Pi}\, A\, \mathrm{e}^{-\mathrm{i}\theta M_\Pi}\,, \tag{7.30}$$

with $(\tau_\theta(A))^* = \tau_\theta(A^*)$ and $\tau_\theta(A_1 \cdot A_2) = \tau_\theta(A_1)\tau_\theta(A_2)$, where A, A_1, A_2 are operators on $\mathscr{H}$, we find, setting

$$\prod_{k=1}^{m} \hat{\Psi}_{j_k}(f_k,\, \alpha_k) =: B\,,$$

and

$$\prod_{l=m+1}^{n} \hat{\Psi}_{j_l}(f_l,\, \alpha_l) =: C\,,$$

that

$$\begin{aligned} \omega(B \cdot C) &= \omega(C\tau_{2\pi\mathrm{i}}(B)) \\ &= \omega(\tau_{-2\pi\mathrm{i}}(C)B)\,, \end{aligned} \tag{7.31}$$

a more familiar form of the KMS condition for $(\omega,\, \tau_\theta)$ at inverse temperature $\beta = 2\pi$; see [62].

It deserves to be noticed that the KMS condition (7.28), (7.31) implies the spin-statistics connection. We calculate formally: For $0 < \varepsilon < \pi$,

$$\begin{aligned} &\omega\big(\hat{\Psi}_{j_1}(f_1,\, 0)\hat{\Psi}_{j_2}(f_2,\, \varepsilon)\big) \\ &\overset{\mathrm{KMS},(7.11)}{=} \mathrm{e}^{-\mathrm{i}2\pi s_{j_2}}\, \omega\big(\hat{\Psi}_{j_2}(f_2,\, 2\pi + \varepsilon)\hat{\Psi}_{j_1}(f_1,\, 0)\big) \\ &\overset{(7.14)}{=} \mathrm{e}^{-\mathrm{i}2\pi s_{j_2}}\, \mathrm{e}^{\mathrm{i}2\pi\theta_{j_1 j_2}}\, \omega\big(\hat{\Psi}_{j_1}(f_1,\, 0)\hat{\Psi}_{j_2}(f_2,\, 2\pi + \varepsilon)\big) \\ &\overset{(7.11)}{=} \mathrm{e}^{-\mathrm{i}2\pi s_{j_2}}\, \mathrm{e}^{\mathrm{i}2\pi\theta_{j_1 j_2}}\, \omega\big(\hat{\Psi}_{j_1}(f_1,\, 0)\hat{\Psi}_{j_2}(f_2,\, \varepsilon)\big)\,. \end{aligned} \tag{7.32}$$

Thus,

$$s_{j_2} = \theta_{j_1\, j_2} \mod \mathbb{Z}\,, \tag{7.33}$$

unless $\omega(\hat{\Psi}_{j_1}(f_1,\, 0)\hat{\Psi}_{j_2}(f_2,\, \varepsilon)) \equiv 0$. If this quantity does *not* vanish (and in $d \geq 3$) then either Ψ_{j_1} and Ψ_{j_2} are both Fermi fields ($\theta_{j_1\, j_2} = \frac{1}{2} \operatorname{mod} \mathbb{Z}$) or they are both Bose fields ($\theta_{j_1\, j_2} = 0 \operatorname{mod} \mathbb{Z}$). Thus, (7.33) proves (a special case of) SSC!

It turns out that the *CPT theorem* (for d *even*) is a direct consequence of the KMS condition (7.31). This claim can be viewed as a corollary of the general Tomita-Takesaki theory [64]. But, in our concrete context, it is easy to directly define an *anti-unitary involution* J acting on $\mathscr{H}$, which, thanks to the KMS condition (7.31), turns out to be a symmetry of the theory: We define

$$B := \hat{\Psi}_{j_1}(f_1, \alpha_1) \cdots \hat{\Psi}_{j_n}(f_n, \alpha_n) , \tag{7.34}$$

with $0 < \alpha_1 < \ldots < \alpha_n < \pi$, and

$$C := \hat{\Psi}_{l_1}(g_1, \beta_1) \cdots \hat{\Psi}_{l_m}(g_m, \beta_m) , \tag{7.35}$$

with $0 < \beta_1 < \ldots < \beta_n < \pi$. We define

$$J B \Omega := \mathrm{e}^{-\pi M_\Pi} B^* \Omega , \tag{7.36}$$

or

$$\begin{aligned} J\hat{\Psi}_{j_1}(f_1, \alpha_1) \cdots \hat{\Psi}_{j_n}(f_n, \alpha_n)\, \Omega \\ &= \mathrm{e}^{-\pi M_\Pi}\, \hat{\Psi}_{\bar{j}_n}(\bar{f}_n, -\alpha_n) \cdots \hat{\Psi}_{\bar{j}_1}(\bar{f}_1, -\alpha_1)\, \Omega \\ &= R^{-1}_{\bar{j}_n}(\pi)\, \hat{\Psi}_{\bar{j}_n}(\bar{f}_n, \pi - \alpha_n) \cdots R^{-1}_{\bar{j}_1}(\pi)\, \hat{\Psi}_{\bar{j}_1}(\bar{f}_1, \pi - \alpha_1)\, \Omega , \end{aligned} \tag{7.37}$$

with $0 < \pi - \alpha_n < \pi - \alpha_{n-1} < \ldots < \pi - \alpha_1 < \pi$. By analytic continuation of (7.37) in the angles $\alpha_1, \ldots, \alpha_n$ to the imaginary axis, see [74], we see that J has the interpretation of the product CP_1T, where P_1 is the space reflection $x^1 \mapsto -x^1$, $\vec{x} \mapsto \vec{x}$, $\vec{x} = (x^2, \ldots, x^{d-1})$; (geometrically, the action of J only involves a reflection in the plane Π). Using (7.36), we find that

$$\begin{aligned} \langle JC\Omega, JB\Omega \rangle &\overset{(7.36)}{=} \langle \mathrm{e}^{-\pi M_\Pi} C^*\Omega, \mathrm{e}^{-\pi M_\Pi} B^*\Omega \rangle \\ &= \langle \mathrm{e}^{-2\pi M_\Pi} C^*\Omega, B^*\Omega \rangle \\ &= \omega(\tau_{-2\pi \mathrm{i}}(C) B^*) \\ &\overset{(7.31)}{=} \omega(B^* C) \\ &= \langle B\Omega, C\Omega \rangle , \end{aligned}$$

which tells us that J is *anti-unitary*. Moreover,

$$\begin{aligned} J(JB\Omega) &= J\left(\mathrm{e}^{-\pi M_\Pi} B^* \mathrm{e}^{\pi M_\Pi}\right) \Omega \\ &= \mathrm{e}^{-\pi M_\Pi} \left(\mathrm{e}^{\pi M_\Pi} B \mathrm{e}^{-\pi M_\Pi}\right) \Omega \\ &= B\Omega , \end{aligned}$$

i.e., J is an *involution.*

In *even* space-time dimension, the product $P_1 P$, where P is space reflection, has determinant $= 1$ and can be represented as a space rotation. Hence $P_1 P$ is a *symmetry* of the theory. It follows that the CPT operator Θ defined by

$$\Theta := J P_1 P \tag{7.38}$$

is an *anti-unitary symmetry* of the theory. This is the celebrated *CPT theorem* [61]. In a space-time of *odd* dimension, the operators $J_j = CP_jT$, $j = 1, \ldots, d-1$ are always anti-unitary symmetries, but, in general, Θ is *not* a symmetry.

For an analysis of SSC and CPT for local RQFT's on a class of curved space-time manifolds with 'large' groups of Killing symmetries (Schwarzschild, de Sitter, AdS), see, e.g., [74].

I conclude my discussion with a result due to *Steven Weinberg* and *Edward Witten*, [75]: In a four-dimensional local RQFT *without gravity*, but with well defined current- and charge density operators, there are no massless charged (asymptotic) particles of spin $> \frac{1}{2}$; and there are no massless (asymptotic) particles of spin > 1 if the theory admits a well defined energy-momentum tensor.

7.2. Braid statistics in two and three space-time dimensions and SSC[15]

Two-dimensional electron gases in a transversal external magnetic field exhibiting the fractional quantum Hall effect appear to be examples of quantum-mechanical systems with fractionally charged quasi-particles having fractional spin $s \notin \frac{1}{2}\mathbb{Z}$ and fractional or braid statistics; see, e.g., [81,82], and references given there. The analysis of these particles is important in order to calculate, e.g., the value of the Hall conductivity σ_H (a rational multiple of $\frac{e^2}{h}$). Certain systems exhibiting the fractional quantum Hall effect (e.g., the ones with $\sigma_H = \frac{5}{2}\frac{e^2}{h}$) are believed to be of interest for purposes of quantum computation. All this is quite fascinating and has been among my more serious scientific interests in the 1990's. Thus, it would have been tempting to give a rather detailed account of the theory of planar systems exhibiting fractional electric charges, fractional spin and fractional or braid statistics.

However, after much agonizing, I have come to the conclusion that it is impossible to give an account of fractional spin and braid statistics that is *accurate* (mathematically precise), *comprehensible*, and *short*. I therefore decided, with considerable regrets, to limit my account of these matters to some very sketchy remarks.

The pure physical states of a quantum-mechanical system with infinitely many degrees of freedom at *zero temperature*, described, e.g., by a local RQFT, fall into different irreducible ('simple') *superselection sectors*. These sectors are invariant under the action of operators corresponding to local observable quantities ('measurements') of the theory. (The action of the algebra of all 'local observables' on every superselection sector of the theory is usually irreducible.) Superpositions of states from different superselection sectors are therefore *incoherent:* Their relative phases are not observable, and interference terms vanish ('decoherence').

---

[15] Sources for this section are [58, 59, 71–73, 76, 79, 80].

Let $I = \{e, 2, 3, \ldots, N\}$, $N \leq \infty$, be a set of indices labeling the different irreducible superselection sectors of such a system, with e labeling the sector containing the *ground state* (or vacuum) Ω of the system. Let U_j, $j \in I$, denote the unitary representation of the quantum-mechanical rotation group $\mathrm{Spin}(d-1)$ on (the Hilbert space $\mathscr{H}_j$ of pure states corresponding to) the superselection sector j. Since the algebra of local observables is assumed to act irreducibly on $\mathscr{H}_j$, and because observables commute with rotations through an angle 2π, one can show that $U_j(R(2\pi))$, where $R(2\pi)$ is a space rotation through an angle 2π, is a multiple of the identity, i.e.,

$$U_j\big(R(2\pi)\big) = \mathrm{e}^{\mathrm{i}2\pi s_j}\, \mathbb{1}_j\,, \tag{7.39}$$

where s_j is called the '*spin (parity) of sector j*'. For $d \geq 4$, $s_j \in \frac{1}{2}\mathbb{Z}$, but, for $d = 3$,

$$\mathrm{Spin}(2) \simeq \mathbb{R}\,, \tag{7.40}$$

so that s_j can, in principle, be an *arbitrary real number* (mod $\mathbb{Z}$).

If the theory describing the system has a *local structure* (see [58, 72, 73, 76]) and the vacuum sector e has appropriate properties ('*Haag duality*', see [58]) then one can show that sectors can be *composed*, i.e., with two sectors, i and j, one can associate their composition, $i \otimes j$, (a kind of *tensor product*), and the sector $i \otimes j$ can be decomposed into a direct sum of irreducible sectors with multiplicities, according to

$$i \otimes j = \bigoplus_{k \in I} N_{ij}^k \cdot k \equiv \bigoplus_{k \in I} \left(\bigoplus_{\alpha=1}^{N_{ij}^k} k^{(\alpha)} \right), \tag{7.41}$$

where $N_{ij}^k = 0, 1, 2, \ldots$ is the *multiplicity* of the irreducible sector k in the tensor product sector $i \otimes j$, and $k^{(\alpha)} \simeq k$. The integers N_{ij}^k are called '*fusion rules*'. If the theory describing the systems has a local structure one can show that:

- $N_{ij}^k = N_{ji}^k$ and $i \otimes j \simeq j \otimes i$;
- to every irreducible sector $j \in I$ one can uniquely associate a (charge-) *conjugate sector* $\bar{j}$ such that $\bar{j} \otimes j \simeq j \otimes \bar{j}$ contains the vacuum (groundstate) sector e, *exactly once*, i.e.,

$$j \otimes \bar{j} = e \oplus \Big(\bigoplus_{\substack{k \in I \\ k \neq e}} N_{ij}^k \cdot k \Big)\,; \tag{7.42}$$

and

- $e \otimes j \simeq j \otimes e \simeq j$, for all $j \in I$.

Since $i \otimes j \simeq j \otimes i$, there must exist an *intertwiner* (morphism) ε_{ij} intertwining $i \otimes j$ with $j \otimes i$:

$$\varepsilon_{ij} \,:\, i \otimes j \xrightarrow{\simeq} j \otimes i\,. \tag{7.43}$$

Focusing on systems in two or three space-time dimensions – which we will do in the following – we find, after some serious reflection, that there are usually *two* distinguished intertwiners ε_{ij}^{+} and ε_{ij}^{-} satisfying (7.43). (In two space-time dimensions, this can be understood to be a consequence of the fact that the complement of a light cone has two disjoint components; in three space-time dimensions, it is related to the circumstance that two points in the plane can be exchanged either clockwise or anti-clockwise.) It turns out that, thanks to the associativity of the composition of sectors (the tensor product $\otimes$), the operators $\varepsilon_{ij}^{\pm}$ obey the *Yang-Baxter equations* (as first observed in [71]), and

$$\varepsilon_{ij}^{+}\,\varepsilon_{ji}^{-} = \text{identity}\,. \tag{7.44}$$

It follows from these properties that the intertwiners $\{\varepsilon_{ij}^{\pm}\,|\, i,j \in I\}$ determine a unitary representation of the groupoid of colored braids on n strands (the colors are the labels of the irreducible sectors, i.e., the elements of I), for arbitrary $n = 2, 3, \ldots$. These representations describe the *quantum statistics* of the system. If

$$\varepsilon_{ij}^{+} = \varepsilon_{ij}^{-} \text{ for all } i,j \in I\,, \tag{7.45}$$

then the representations of the braid groupoids are actually representations of the *permutation groups*, and the quantum statistics ultimately reduces to ordinary *Bose / Fermi statistics.* In $d \geq 4$ space-time dimensions, Eq. (7.45) always holds.

Let $\mathbb{N}_i$ denote the $|I| \times |I|$ matrix with positive integer matrix elements

$$(\mathbb{N}_i)_j^k = N_{ij}^k\,. \tag{7.46}$$

The matrices $\mathbb{N}_i$, $i \in I$, all commute and have a common Perron-Frobenius eigenvector Δ, with components $\Delta_i \geq 0$, $i \in I$. It is quite easy to show, using (7.41) - (7.43), that

$$\mathbb{N}_i\,\Delta = \Delta_i\,\Delta\,, \tag{7.47}$$

i.e., Δ_i is the largest eigenvalue of the matrix $\mathbb{N}_i$; Δ_i, is called the *statistical* (or *quantum*) *dimension* of the sector i. Clearly $\mathbb{N}_e = \mathbb{1}$ and hence $\Delta_e = 1$. If *all* statistical dimensions Δ_i, $i \in I$, are *positive integers* then the quantum statistics is ordinary Bose / Fermi statistics or *abelian* braid statistics. Thus *non-abelian* braid statistics is only encountered in theories with some fractional quantum dimensions.

Next, we introduce the '*monodromy operators*'

$$\mu_{ij} := \varepsilon_{ij}^{+}\,\varepsilon_{ji}^{+}\,. \tag{7.48}$$

One aspect of the general *connection between spin and statistics* is that the spectrum of the monodromy operator μ_{ij} consists of the eigenvalues

$$\exp\left[\mathrm{i}2\pi(s_i + s_j - s_k)\right],\ k \in I\,, \tag{7.49}$$

and the multiplicity of the eigenvalue $\exp\left[\mathrm{i}2\pi(s_i + s_j - s_k)\right]$ is given by N_{ij}^k; see [73, 79]. Let v_{ij}^k be an intertwiner ('Clebsch-Gordan operator') intertwining the sector $i \otimes j$ with a subsector k; see (7.43). There are precisely N_{ij}^k *linearly independent* such intertwiners.

Then

$$\mu_{ij}\,v_{ij}^k = \exp\left[\mathrm{i}2\pi(s_i + s_j - s_k)\right] v_{ij}^k\,. \tag{7.50}$$

In particular, for $i = \bar{j}$, $k = e$, we have that

$$\mu_{\bar{j}j}\, v^{e}_{\bar{j}j} = \exp\left[\mathrm{i}2\pi(s_j + s_{\bar{j}})\right] v^{e}_{\bar{j}j}\,, \tag{7.51}$$

because $s_e = 0 \bmod \mathbb{Z}$. One can show that

$$s_j = -s_{\bar{j}} \bmod \mathbb{Z}\,, \tag{7.52}$$

or, equivalently,

$$\mu_{\bar{j}j}\, v^{e}_{\bar{j}j} = v^{e}_{\bar{j}j}\,.$$

This is a weaker form of Eq. (7.22), subsection 7.1. We conclude this brief survey with the following result valid (for local RQFT in) *three* space-time dimensions and established in [73]; (see also references given there).

Theorem 7.1.

(1) *If I is a finite set then s_j is a* rational number, *for all $j \in I$.*

(2) *If either* space reflection in a line *or* time reversal *is a* symmetry *of the theory on* all *its superselection sectors $j \in I$ then the quantum statistics of the theory is ordinary* permutation-group *(Bose / Fermi)* statistics, *and*

$$s_j \in \frac{1}{2}\mathbb{Z}\,, \text{ for all } j \in I\,. \tag{7.53}$$

(3) *The following two statements are* equivalent*:*

  (i) *The quantum statistics of the theory is ordinary permutation-group (Bose/Fermi) statistics.*

  (ii) $\exp\left[\mathrm{i}2\pi(s_i + s_j - s_k)\right] = 1$, *for all i, j, k in I with $N^k_{ij} \geq 1$.*

Moreover, both statements imply that

$$s_j \in \frac{1}{2}\mathbb{Z}\,, \text{ for all } j \in I\,.$$

Remarks.

(1) The *rationality* of the Hall conductivity, i.e., $\sigma_H = r\frac{e^2}{h}$, $r \in \mathbb{Q}$, in two-dimensional, incompressible electron gases exhibiting the fractional quantum Hall effect is intimately connected to part *(1)* of the theorem; see [82].

(2) Space reflections in a line and time reversal are *not* symmetries of a two-dimensional electron gas in a transversal, external magnetic field. In view of part *(2)* of the theorem, this explains why such systems may exhibit quasi-particles with braid statistics.

(3) The precise hypotheses under which the theorem is proven (e.g., local RQFT satisfying 'Haag duality') can be found in [73].

It is not entirely easy to translate the contents of this theorem into purely field theoretic jargon, at least if one desires to be precise, mathematically. The remark may help the reader that 'physical' examples of sectors with fractional spin and braid statistics can be found in the realm of abelian and non-abelian Chern-Simons theories; see, e.g., [70, 83]. In these theories, sectors with fractional spin and statistics can be constructed by applying field operators with Mandelstam flux strings to the vacuum sector. In the theory of the quantum Hall effect topological versions of these theories play a fundamental rôle; see [82]. They also appear in the theoretical description of graphene.

Well, I guess it is time to claim victory!

References

[1] A. Pais, *Inward Bound*, Oxford University Press, New York, 1986.

[2] H. Kragh, *Quantum Generations*, Princeton University Press, Princeton, 1999.

[3] N. Straumann, *Über Paulis wichtigste Beiträge zur Physik*, preprint, arXiv:physics/001003.

[4] D. Giulini, *Electron Spin or 'Classically Non-Describable Two-Valuedness'*, preprint, arXiv:hist-ph/0710.3128.

[5] J. Fröhlich, *Réflexions sur Wolfgang Pauli*, proceedings of the "Colloque 2000: Pensée et Science" of the Fondation F. Gonseth, Eric Emery (ed.), Rev. Synt. **126** (2005), 443–450.

[6] N. Straumann, *Quantenmechanik*, Springer-Verlag, 2002.

[7] W. Pauli, Z. Physik **16** (1923), 155–164.

[8] W. Pauli, Z. Physik **31** (1925), 373–385.

[9] S. Ferrara, M. Porrati, V.L. Telegdi, Phys. Rev. D **46** (1992), 3529–3537.

[10] J.D. Jackson, *Classical Eletromagnetism*, John Wiley & Sons, New York, 1975.

[11] W. Pauli, Z. Physik **31** (1925), 765–783.

[12] E.C. Stoner, Phil. Magazine **48** (1924), 719–736.

[13] W. Heisenberg, Zeitschrift für Physik **33** (1925), 879–893.

[14] M. Born, P. Jordan, Zeitschrift für Physik **34** (1925), 858–888.

[15] M. Born, W. Heisenberg, P. Jordan, Zeitschrift für Physik **35** (1926), 557–615.

[16] P.A.M. Dirac, Proc. Royal Soc. (London) A **109** (1925), 642–653.

[17] E. Schrödinger, Annalen der Physik **79** (1926), 361–376; Annalen der Physik **76** (1926), 146–147; Annalen der Physik **80** (1926), 437–490.

[18] P.A.M. Dirac, Proc. Royal Soc. A **117** (1928), 610; A **118** (1928), 351.

[19] W. Pauli, Z. Physik **43** (1927), 601–623.

[20] J. Fröhlich, O. Grandjean, A. Recknagel, Comm. Math. Phys. **193** (1998), 527–594.

[21] J. Fröhlich, *The Electron is Inexhaustible*, Amer. Math. Soc. Publ., Providence RI, 1999.

[22] D. Salamon, *Spin Geometry and Seiberg-Witten Invariants*, preprint 1995.

[23] A. Connes, *Noncommutative Geometry*, Academic Press, New York 1994.

[24] J. Fuchs, Chr. Schweigert, *Symmetries, Lie Algebras and Representations*, Cambridge University Press, Cambridge, New York 1997.

[25] G. Velo, D. Zwanziger, Phys. Rev. **186** (1969), 1337–1341; Phys. Rev. **188** (1969), 2218–2222;

A.Z. Capri, R.L. Kobes, Phys. Rev. D **22** (1980), 1967–1978.

[26] S. Deser, B. Zumino, Phys. Lett. B **62** (1976), 335;

K. Vonlanthen, *Supergravitation und Velo-Zwanziger Phänomene*, ETH diploma thesis 1978 (N. Straumann, advisor).

[27] J. Fröhlich, U.M. Studer, E. Thiran, *Quantum Theory of Large Systems of Non-Relativistic Matter*, in: *Fluctuating Geometries in Statistical Mechanics and Field Theory*, Les Houches, Session LXII (1994), F. David, P. Ginsparg, J. Zinn-Justin (eds.), Elsevier, New York, 1996.

[28] R. Howe, Lect. Appl. Math. **21** (1985), 179.

[29] S. Weinberg, *The Quantum Theory of Fields*, Vol. 1, Cambridge University Press, Cambridge, New York, 1995;

J. Fröhlich, *Einführung in die Quantenfeldtheorie*, ETH Lecture Notes, 1986.

[30] W. Hunziker, Commun. Math. Phys. **40** (1975), 215–222.

[31] E.H. Lieb, *The Stability of Matter: From Atoms to Stars*, 4th edition, Springer-Verlag, 2005.

[32] J. Fröhlich, E.H. Lieb, M. Loss, Commun. Math. Phys. **104** (1986), 251–270.

[33] M. Loss, H.-T. Yau, Commun. Math. Phys. **104** (1986), 283–290.

[34] E.H. Lieb, M. Loss, Commun. Math. Phys. **104** (1986), 271–282.

[35] C. Fefferman, Proc. Natl. Acad. Science USA **92** (1985), 5006–5007, and Lecture Notes.

[36] E.H. Lieb, M. Loss, J.-Ph. Solovej, Phys. Rev. Letters **75** (1995), 985–989.

[37] J. Fröhlich, Ann. Inst. H. Poincaré **19** (1974), 1–103; Fortschritte der Physik **22** (1974), 159–198.

[38] L. Bugliaro Goggia, J. Fröhlich, G.M. Graf, Phys. Rev. Letters **77** (1996), 3494–3497.

[39] C. Fefferman, J. Fröhlich, G.M. Graf, Proc. Natl. Acad. Sci. **93** (1996), 15009–15011.

[40] C. Fefferman, J. Fröhlich, G.M. Graf, Commun. Math. Phys. **190** (1999), 309–330.

[41] L. Bugliaro Goggia, C. Fefferman, J. Fröhlich, G.M. Graf, J. Stubbe, Commun. Math. Phys. **187** (1997), 567–582.

[42] L. Bugliaro Goggia, C. Fefferman, G.M. Graf, Revista Matematica Iberoamericana **15** (1999), 593–619.

[43] V. Bach, J. Fröhlich, I.M. Sigal, Adv. Math. **137** (1998), 205–298; **137** (1998), 299–395.

[44] V. Bach, J. Fröhlich, I.M. Sigal, Commun. Math. Phys. **207** (1999), 249–290.

[45] M. Griesemer, M. Loss, E.H. Lieb, Inventiones Math. **145** (1999), 557–587.

[46] J. Fröhlich, M. Griesemer, B. Schlein, Adv. Math. **164** (2001), 349–398.

[47] J. Fröhlich, M. Griesemer, B. Schlein, Ann. Henri Poincaré **3** (2002), 107–170.

[48] V. Bach, J. Fröhlich, A. Pizzo, Comm. Math. Phys. **264** (2006), 145–165; Comm. Math. Phys. **274** (2007), 457–486; Adv. Math. (to appear).

[49] T. Chen, J. Fröhlich, A. Pizzo, *Infraparticle Scattering States in Non-Relativistic QED: I & II*, preprints 2007.

[50] H. Spohn, *Dynamics of Charged Particles and Their Radiation Field*, Cambridge University Press, Cambridge, New York, 2004.

[51] T. Chen, ETH Diploma Thesis, 1994.

[52] R. Stora, Private communication.

[53] G. Gabrielse et al.; see Gabrielse's contribution in this book.

[54] M. Fierz, Helv. Phys. Acta **12** (1939), 3.

[55] R. Jost, *The General Theory of Quantized Fields*, AMS Publ., Providence RI, 1965.

[56] R.F. Streater, A.S. Wightman, *PCT, Spin and Statistics and All That*, Benjamin, New York, 1964.

[57] J. Glimm, A. Jaffe, *Quantum Physics: A functional Integral Point of View*, Springer-Verlag, 1987.

[58] S. Doplicher, R. Haag, J.E. Roberts, Commun. Math. Phys. **33** (1971), 199; Commun. Math. Phys. **35** (1974), 49.

[59] S. Doplicher, J.E. Roberts, Commun. Math. Phys. **131** (1990), 51.

[60] G. Lüders, Kong. Dansk. Vid. Selskab, Mat.-Fys. Medd. **28** (1954), 5; Ann. Phys. **2** (1957), 1;
W. Pauli, Nuovo Cimento **6** (1957), 204.

[61] R. Jost, Helv. Phys. Acta **30** (1957), 409.

[62] R. Kubo, J. Phys. Soc. Japan **12** (1957), 57;
P.C. Martin, J. Schwinger, Phys. Rev. **115** (1959), 1342;
R. Haag, N. Hugenholtz, M. Winnink, Commun. Math. Phys. **5** (1967), 215.

[63] J.J. Bisognano, E.H. Wichmann, J. Math. Phys. **16** (1975), 985–1007.

[64] M. Takesaki, *Tomita's Theory of Modular Hilbert Algebras and its Applications*, Lecture Notes in Mathematics **128**, Springer-Verlag, 1970;
O. Bratteli, D.W. Robinson, *Operator Algebras and Quantum Statistical Mechanics*, Springer-Verlag, 1979, 1981.

[65] R.F. Streater, I.F. Wilde, Nucl. Phys. B **24** (1970), 561.

[66] J. Fröhlich, Commun. Math. Phys. **47** (1976), 269–310.

[67] M. Leinaas, J. Myrheim, Il Nuovo Cimento **37 B** (1977), 1.

[68] G.A. Goldin, R. Menikoff, D.H. Sharp, J. Math. Phys. **22** (1981), 1664.

[69] F. Wilczeck, Phys. Rev. Letters **48** (1982), 1144; **49** (1982), 957.

[70] J. Fröhlich, P.A. Marchetti, Lett. Math. Phys. **16** (1988), 347; Commun. Math. Phys. **121** (1988), 177.

[71] J. Fröhlich, *Statistics of Fields, the Yang-Baxter Equation and the Theory of Knots and Links*, in: *Non-Perturbative Quantum Field Theory*, Cargèse 1987, G. 't Hooft et al. (eds.), Plenum Press, New York, 1988.
J. Fröhlich, *Statistics and Monodromy in Two- and Three-Dimensional Quantum Field Theory*, in: *Differential Geometrical Methods in Theoretical Physics*, K. Bleuler, M. Werner (eds.), Kluwer Academic Publ., Dordrecht, 1988.

[72] K. Fredenhagen, K.H. Rehren, B. Schroer, Commun. Math. Phys. **125** (1989), 201.

[73] J. Fröhlich, F. Gabbiani, Rev. Math. Phys. **2** (1990), 251;
J. Fröhlich, P.A. Marchetti, Nucl. Phys. B **356** (1991), 533.

[74] L. Birke, J. Fröhlich, Rev. Math. Phys. **14** (2002), 829.

[75] S. Weinberg, E. Witten, Phys. Letters B **96** (1980), 59.

[76] D. Buchholz, K. Fredenhagen, Commun. Math. Phys. **84** (1982), 1.

[77] K. Osterwalder, R. Schrader, Commun. Math. Phys. **42** (1975), 281; see also V. Glaser, Commun. Math. Phys. **37** (1974), 257.

[78] H. Araki, J. Math. Phys. **2** (1961), 267;
W. Schneider, Helv. Phys. Acta **42** (1969), 201.

[79] J. Fröhlich, T. Kerler, *Quantum Groups, Quantum Categories and Quantum Field Theory*, Lecture Notes in Mathematics, Vol. 1542, Springer-Verlag, 1993.

[80] J. Fuchs, I. Runkel, Chr. Schweigert, Nucl. Phys. B **624** (2002), 452; Nucl. Phys. B **646** (2002), 353.

[81] R.E. Prange, S.M. Girvin (eds.), *The Quantum Hall Effect*, Graduate Texts in Contemporary Physics, Springer-Verlag, 1990;
M. Stone (ed.), *Quantum Hall Effect*, World Scientific Publ., Singapore, London, Hong Kong, 1992.

[82] J. Fröhlich, *The Fractional Quantum Hall Effect, Chern-Simons Theory, and Integral Lattices*, in: Proc. of ICM '94, S.D. Chatterji (ed.), Birkhäuser Verlag, 1995.
J. Fröhlich, B. Pedrini, Chr. Schweigert, J. Walcher, J. Stat. Phys. **103** (2001) 527;
J. Fröhlich, B. Pedrini, in: *Statistical Field Theory*, Como 2001, A. Cappelli, G. Mussardo (eds.), Kluwer, New York, Amsterdam, 2002.

[83] R. Jackiw, S. Templeton, Phys. Rev. D **23** (1981), 2291;
S. Deser, R. Jackiw, S. Templeton, Phys. Rev. Letters **48** (1982), 975;
R. Pisarski, S. Rao, Phys. Rev. D **32** (1985), 2081.

Jürg Fröhlich
Theoretical Physics
ETH Zürich
CH–8093 Zürich
Louis-Michel visiting professor at IHÉS
e-mail: `juerg@itp.phys.ethz.ch`

The Spin
B. Duplantier, J.-M. Raimond and V. Rivasseau, Eds., 61–69

Poincaré Seminar 2007

New Kinds of Quantum Statistics

Frank Wilczek

Abstract. I review the quantum kinematics of identical particles, which suggests new possibilities, beyond bosons and fermions, in 2+1 dimensions; and how simple flux-charge constructions embody the new possibilities, leading to both abelian and nonabelian anyons. I briefly allude to experimental realizations, and also advertise a spinor construction of nonabelian statistics, that has a 3+1 dimensional extension.

In quantum theory the notion of identity reaches a new level of precision and has profound dynamical significance. It becomes important that two particles can be *precisely* identical, i.e. indistinguishable, as opposed to merely similar. When passing from a classical description of indistinguishable particles to a quantum description one must supply additional rules, known as the quantum statistics of the particles.

For many years it was thought that there are only two possibilities for quantum statistics: bosons and fermions. But in 1977 Leinaas and Myrheim [1] demonstrated, at the level of particle quantum mechanics, that there were additional theoretically consistent possibilities. I'm told there were hints of this in earlier work in axiomatic field theory [2]. The work of Leinaas and Myrheim received little attention, and their insight was rediscovered, independently, by Goldin, Menikoff, and Sharp [3], who realized it in the context of a special formulation of quantum mechanics using currents and densities; and by me using conventional quantum field theory [4] (involving solitons, as below, and/or Chern-Simons terms [5]). This circle of ideas came to life as physics in 1984, when Arovas, Schrieffer and I demonstrated [6] – theoretically, but I think quite convincingly – that quasiparticles in the fractional quantum Hall effect obey forms of the new, "anyon" quantum statistics. (That possibility was foreseen by Halperin [7].). The anyonic behavior of quasiparticles (and quasiholes) in the fractional quantum Hall effect is so closely integrated into the overall theory of those states that it can be subtle to demonstrate as an independent phenomenon. A recent series of impressive experiments by V. Goldman and his collaborators [8] have been interpreted this way, and other experiments, requiring less interpretation, are in the works.

Rich mathematical possibilities arise when we consider *nonabelian* statistics. In the abelian case the operations characteristic of quantum statistics – roughly speaking: slow, distant exchange of particle positions – are implemented as multiplications of the wave-function by a complex number (phase). In the nonabelian case complex motions in large Hilbert spaces of degenerate states can come into play. The possibility of exploiting a robust mapping from operations in physical space (characterized topologically) to navigate through large Hilbert spaces has inspired visions of a possible route to quantum computing, known as topological quantum computing. Physical realization of topological quantum computing is still far off, if it can be achieved at all, but the program has inspired impressive work, both theoretical and experimental. An upcoming milestone may be demonstration of a proposal by Moore and Read [9] that quasiparticles in an observed $\nu = \frac{5}{2}$ quantum Hall state obey nonabelian statistics. Experimental programs to test this are well advanced, as well.

Here I will describe a few of the most fundamental concepts underlying these developments in what might appear, to a quantum field theorist, as their simplest natural context. (I will mention quantum Hall physics, experimental aspects, and quantum computing, but I will not even begin to do them justice.) In the course of this review a few intriguing new ideas will come up, too.

1. Braids, Permutations, and In Between

Traditionally, the world has been divided between bosons (Bose-Einstein statistics) and fermions (Fermi-Dirac statistics). Let's recall what these are, and why they appear to exhaust the possibilities.

If two identical particles start at positions (A, B) and transition to (A', B'), we must consider both $(A, B) \rightarrow (A', B')$ and $(A, B) \rightarrow (B', A')$ as possible accounts of what has happened. According the rules of quantum mechanics, we must add the amplitudes for these possibilities, with appropriate weights. The rules for the weights encode the dynamics of the particular particles involved, and a large part of what we do in fundamental physics is to determine such rules and derive their consequences.

In general, discovering the rules involves creative guesswork, guided by experiment. One important guiding principle is correspondence with classical mechanics. If we have a classical Lagrangian $L_{\text{cl.}}$, we can use it, following Feynman, to construct a path integral, with each path weighted by a factor

$$e^{i \int dt L_{\text{cl.}}} \equiv e^{i S_{\text{cl.}}} \tag{1}$$

where $S_{\text{cl.}}$ is the classical action. This path integral provides – modulo several technicalities and qualifications – amplitudes that automatically implement the general rules of quantum mechanics. Specifically: it sums over alternative histories, takes products of amplitudes for successive events, and generates unitary time evolution.

The classical correspondence, however, does not instruct us regarding the relative weights for trajectories that are topologically distinct, i.e. that cannot be continuously deformed into one another. Since only small variations in trajectories are involved in determining the classical equations of motion, from the condition that $S_{\text{cl.}}$ is stationary, the classical equations cannot tell us how to interpolate between topologically distinct trajectories. We need additional, essentially quantum-mechanical rules for that.

Now trajectories that transition $(A, B) \rightarrow (A', B')$ respectively $(A, B) \rightarrow (B', A')$ are obviously topologically distinct. The traditional additional rule is: for bosons, add the amplitudes for these two classes of trajectories[1]; for fermions, subtract.

These might appear to be the only two possibilities, according to the following (not-quite-right) argument. Let us focus on the case $A = A', B = B'$. If we run an "exchange" trajectory $(A, B) \rightarrow (B, A)$ twice in succession, the doubled trajectory is a direct trajectory. The the square of the factor we assign to the exchange trajectory must be the square of the (trivial) factor 1 we associate to the direct trajectory, i.e. it must be ± 1.

This argument is not conclusive, however, because there can be additional topological distinctions among trajectories, not visible in the mapping between endpoints. This distinction is especially important in 2 spatial dimensions, so let us start there. (I should recall that quantum-mechanical systems at low energy can effectively embody reduced dimensionality, if their dynamics is constrained below an energy gap to exclude excited states whose wave functions have structure in the transverse direction.) The topology of trajectory space is then specified by the *braid group*. Suppose that we have N identical particles. Define the elementary operation σ_j to be the act of taking particle j over particle $j + 1$, so that their final positions are interchanged, while leaving the other particles in place. (See Figure 1.) We define products of the elementary operations by performing them sequentially. Then we have the obvious relation

$$\sigma_j \sigma_k = \sigma_k \sigma_j ; \quad |j - k| \geq 2 \tag{2}$$

among operations that involve separate pairs of particles. We also have the less obvious *Yang-Baxter* relation

$$\sigma_j \sigma_{j+1} \sigma_j = \sigma_{j+1} \sigma_j \sigma_{j+1} \tag{3}$$

which is illustrated in Figure 1. The topologically distinct classes of trajectories are constructed by taking products of σ_js and their inverses, subject only to these relations.

If we add to the relations that define the braid group the additional relations

$$\sigma_j^2 = 1 \tag{4}$$

then we arrive at the symmetric (permutation) group S_N. In 3 spatial dimensions, there are more ways to untangle trajectories. Indeed, one can always untangle two

[1] As determined by the classical correspondence, or other knowledge of the interactions.

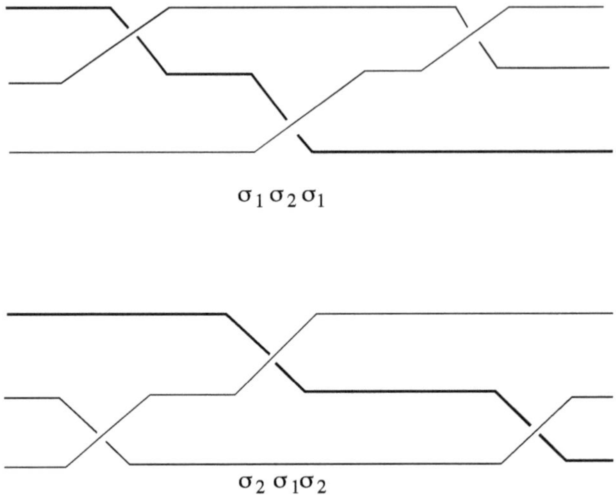

FIGURE 1. The elementary acts of crossing one particle trajectory over another generate the braid group. The Yang-Baxter relation $\sigma_1\sigma_2\sigma_1 = \sigma_2\sigma_1\sigma_2$, made visible here, is its characteristic constraint.

world-lines by escaping into the transverse direction to avoid potential intersections, so the permutation of endpoints captures all the topology.

Yet in 3 dimensions, famously, rotations through 2π are not topologically trivial. This topological fact underlies the possibility of spin-$\frac{1}{2}$ (projective) representations of the rotation group. In such representations, the action of a 2π rotation is to multiply the wave function by -1. On the other hand, rotations through 4π are topologically trivial. This suggests that for particles with extended structure, that cannot be adequately represented as simple points (e.g., magnetic monopoles, or solitons with extended zero-modes) we should consider relaxing Eqn. (4) to

$$\sigma_j^4 = 1 \tag{5}$$

since σ_j^2 can be implemented by a 2π rotation moving the particles $(j, j+1)$ around one another, and σ_j^4 by a 4π rotation. The relations Eqn. (5), together with Eqns. (2, 3), define a group intermediate between the braid group and the symmetric group.

2. Abelian Anyons

The substitution

$$\sigma_j \rightarrow e^{i\theta} \tag{6}$$

preserves the defining relations of the braid group for any phase factor $e^{i\theta}$, so it generates a unitary representation of the braid group. Thus, at the level of quantum kinematics, it is consistent to weight the amplitudes from topologically distinct classes of trajectories with the corresponding phase factors. (Of course, the additional constraint Eqn. (4) reduces the freedom to $e^{i\theta} = \pm 1$.) This possibility defines the classic, abelian anyons.

There is a simple dynamical realization of anyons, using flux and charge. Consider a $U(1)$ gauge theory that has particles of charge q, associated with a field η and is spontaneously broken by a condensate associated with a field of ρ of charge mq, with m an integer[2] Gauge transformations that multiply η by $e^{2\pi ik/m}$ will multiply ρ by $e^{2\pi ik}$. Thus for integer k they will leave the condensate invariant, but generally act nontrivially on η. We are left with an unbroken gauge group Z_m, the integers modulo m. No conventional long-range gauge interaction survives the symmetry breaking, but there is a topological interaction, as follows:

The theory supports vortices with flux quantized in units of

$$\Phi_0 = \frac{2\pi}{mq} \tag{7}$$

in units with $\hbar \equiv 1$. A particle or group of particles with charge bq moving around a flux Φ will acquire a phase

$$\exp i(\oint dt\vec{v}\cdot\vec{A}) = \exp i(\oint d\vec{x}\cdot\vec{A}) = e^{i\Phi bq}. \tag{8}$$

If the flux is $b\Phi_0$, then the phase will be $e^{2\pi i\frac{ab}{m}}$.

Composites with (flux, charge) = $(a\Phi_0, bq)$ will be generally be anyons: as we implement the interchange σ_j, each charge cluster feels the influence of the other's flux. (Note that in two dimensions the familiar flux tubes of three-dimensional physics degenerate to points, so it is proper to regard them as particles.) There are also topological interactions, involving similar accumulations of phase, for non-identical particles. What matters are the quantum numbers, or more formally the superselection sector, not the detailed structure of the particles or excitations involved.

The phase factors that accompany winding have observable consequences. They lead to a characteristic "long range" contribution to the scattering cross-section[3], first computed by Aharonov and Böhm [10] in their classic paper on the significance of the vector potential in quantum mechanics. Unfortunately, that cross-section may not be easy to access experimentally for anyons that occur as excitations in exotic states of condensed matter.

Interferometry appears more practical. The basic concept is simple and familiar, both from optics and (for instance) from SQUID magnetometers. One divides a coherent flow into two streams, which follow different paths before recombining. The relative phase between the paths determines the form of the interference, which can range from constructive to destructive recombination of the currents.

---

[2]If m is irrational the gauge group is not compact, i.e. it is the additive group $\mathbf{R}^+$ rather than $U(1)$.

[3]It diverges at small momentum transfer and in the forward direction.

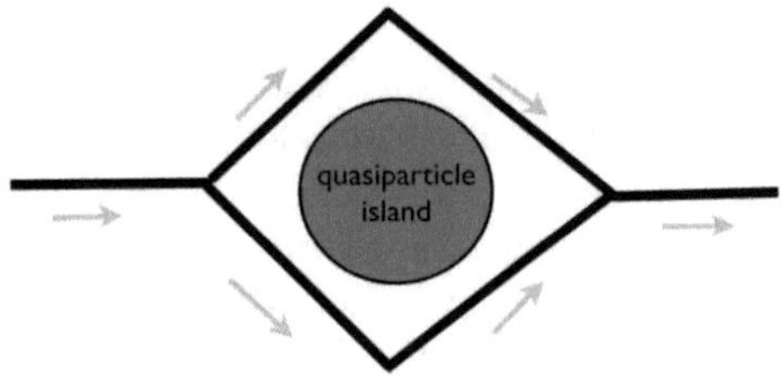

FIGURE 2. A schematic interference experiment to reveal quantum statistics. We study how the combined current depends on the occupation of the quasiparticle island.

We can vary the superselection sector of the area bounded by the paths, and look for corresponding, characteristic changes in the interference. (See Figure 2.) Though there are many additional refinements, this is the basic concept behind both the Goldman experiments and other planned anyon detection experiments [11].

Elementary excitations in the fractional quantum Hall effect are predicted to be anyons. By far the simplest states to analyze are the original Laughlin $1/m$ states, where the excitations are anyons with $\theta = \pi/m$. There is a rich theory covering more general cases.

3. Nonabelian Anyons

The preceding field-theoretic setting for abelian anyons immediately invites nonabelian generalization. We can have a nonabelian gauge theory broken down to a discrete nonabelian subgroup; vortex-charge composites will then exhibit long range, topological interactions of the same kind as we found in the abelian case, for the same reason.

The mathematics and physics of the nonabelian case is considerably more complicated than the abelian case, and includes several qualitatively new effects. First, and most profoundly, we will find ourselves dealing with irreducible *multidimensional* representations of the braiding operations. Thus by winding well-separated particles[4] around one another, in principle arbitrarily slowly, we can not

[4]From here on I will refer to the excitations simply as particles, though they may be complex collective excitations in terms of the underlying electrons, or other degrees of freedom.

only acquire phase, but even navigate around a multidimensional Hilbert space. For states involving several particles, the size of the Hilbert spaces can get quite large: roughly speaking, they grow exponentially in the number of particles.

As will appear, the states in question are related by locally trivial but globally non-trivial gauge transformations. Thus they should be very nearly degenerate. This situation is reminiscent of what one would have if the particles had an internal of freedom – a spin, say. However the degrees of freedom here are not localized on the particles, but more subtle and globally distributed.

The prospect of having very large Hilbert spaces that we can navigate in a controlled way using topologically defined (and thus forgiving!), gentle operations in physical space, and whose states differ in global properties not easily obscured by local perturbations, has inspired visions of *topological quantum computing.* (Preskill [12] has written an excellent introductory review.) The journey from this vision to the level of engineering practice will be challenging, to say the least, but thankfully there are fascinating prospects along the way.

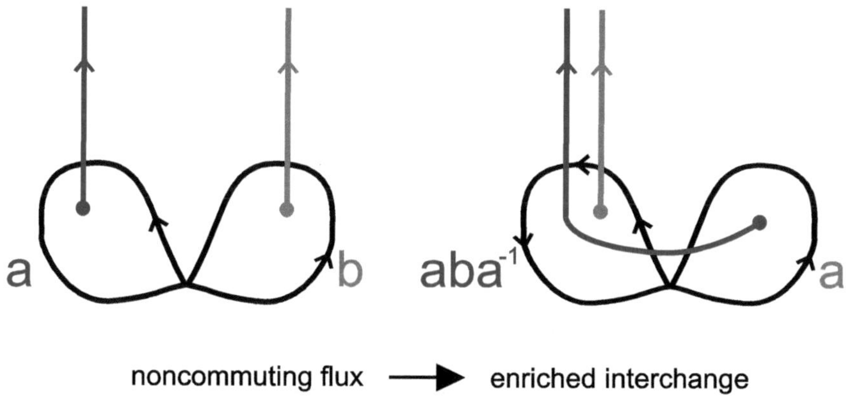

FIGURE 3. By a gauge transformation, the vector potential emanating from a flux point can be bundled into a singular line. This aids in visualizing the effects of particle interchanges. Here we see how nonabelian fluxes, as measured by their action on standardized particle trajectories, are modified by particle interchange.

The tiny seed from which all this complexity grows is the phenomenon displayed in Figure 3. To keep track of the topological interactions, it is sufficient to know the total (ordered) line integral of the vector potential around simple circuits issuing from a fixed base point. This will tell us the group element a that will be applied to a charged particle as it traverses that loop. (The value of a generally depends on the base point and on the topology of how the loop winds around the regions where flux is concentrated, but not on other details. More formally, it gives a representation of the fundamental group of the plane with punctures.) If a charge that belongs to the representation R traverses the loop, it will be transformed according to $R(a)$. With these understandings, what Figure 3 makes clear is that when two flux points with flux (a, b) get interchanged by winding the second over the first, the new configuration is characterized as (aba^{-1}, a). Note here that we cannot simply pull the "Dirac strings" where flux is taken off through one another, since nonabelian gauge fields self-interact! So motion of flux tubes in physical space generates non-trivial motion in group space, and thus in the Hilbert space of states with group-theoretic labels.

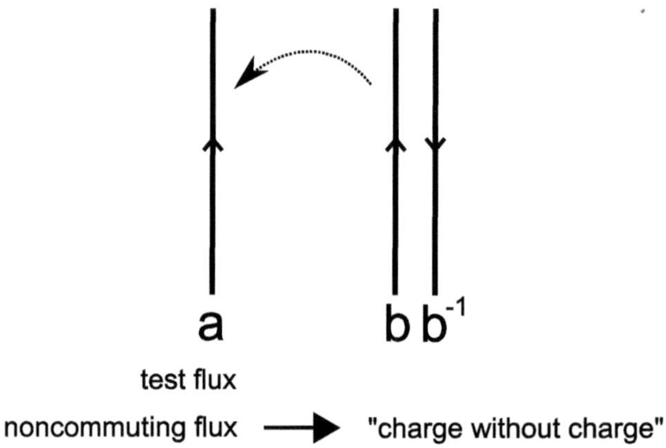

FIGURE 4. Winding a flux-antiflux pair around a test flux, and seeing that it gets conjugated, we find that the pair carries charge.

As a small taste of the interesting things that occur, consider the slightly more complicated situation displayed in Figure 4, with a pair of fluxes (b, b^{-1}) on the right. It's a fun exercise to apply the rule for looping repeatedly, to find out what happens when we take this pair all the way around a on the right. One finds

$$(a, (b, b^{-1})) \rightarrow (a, (aba^{-1}, ab^{-1}a^{-1})) \tag{9}$$

i.e., the pair generally has turned into a different (conjugated) pair. Iterating, we eventually close on a finite-dimensional space of different kinds of pairs. There is a non-trival transformation $\tilde{R}(a)$ in this space that implements the effect of the flux a on pairs that wind around it. But this property – to be transformed by the group operation – is the defining property of charge! We conclude that flux pairs – flux and inverse flux – act as charges. We have constructed, as John Wheeler might have said, Charge Without Charge.

This flux construction makes it clear that nonabelian statistics is consistent with all the general principles of quantum field theory. Physical realization in condensed matter is a different issue – in that context, nonabelian gauge fields don't come readily to hand. Fortunately, and remarkably, there may be other ways to get there. At least one state of the quantum Hall effect, the so-called

Moore-Read state at filling fraction $\frac{5}{2}$, has been identified as a likely candidate to support excitations with nonabelian statistics.

The nonabelian statistics of the Moore-Read state is closely tied up with spinors [13] [14]. I'll give a proper discussion of this, including an extension to 3 + 1 dimensions, elsewhere [15]. Here, I'll just skip to the chase. Taking N γ_j matrices satisfying the usual Clifford algebra relations

$$\{\gamma_j, \gamma_k\} = 2\delta_{jk} \tag{10}$$

the braiding σ_j are realized as

$$\sigma_j = e^{i\pi/4}\frac{1}{\sqrt{2}}(1 + \gamma_j\gamma_{j+1}). \tag{11}$$

It's an easy exercise to show that these obey Eqns. (2, 3), and $\sigma_j^4 = 1$ (Eqn. (5)) but not $\sigma_j^2 = 1$ (Eqn. (4)).

References

[1] J. M. Leinaas and J. Myrheim, Nuovo Cimento **B37** (1977), 1 .

[2] R. Doplicher, R. Haag, and J. Roberts, Comm. Math. Phys. **35** (1972), 49.

[3] G. Goldin, R. Menikoff, and D. Sharp J. Math. Phys. **22** (1981), 1664.

[4] F. Wilczek, Phys. Rev. Lett. **48** (1982), 1144; ibid. (1982), 1146; ibid. **49**(1982), 957.

[5] F. Wilczek and A. Zee, Phys. Rev. Lett. **51** (1983), 2250; D. Arovas, J. R. Schrieffer, F. Wilczek, and A Zee Nucl. Phys. **B251** [FS 13] (1985), 917.

[6] D. Arovas, J. R. Schrieffer and F. Wilczek, Phys. Rev. Lett. **53** (1984), 722.

[7] B. Halperin, Phys. Rev. Lett. **52** (1984), 1583.

[8] V. Goldman, J. Liu and A. Zaslavsky, Phys. Rev. **B71** (2005), 153303; F. Camino, F. Zhou, and V. Goldman Phys. Rev. Lett. **98** (2007), 076805.

[9] G. Moore and N. Read, Nucl. Phys. **B360** (1991), 362.

[10] Y. Aharonov and D. Bohm, Phys. Rev. **115** (1959), 485.

[11] See M. Dolev, M. Heiblum, V. Umansky, A. Stern, and D. Mahalu, Nature **452** (2008), 829; I. Radu, J. Miller, C. Marcus, M. Kastner, L. Pfeiffer, and K. West, Science **320** (2008), 899.

[12] J. Preskill, www.theory.caltech.edu/preskill/ph219 (2004-)

[13] C. Nayak and F. Wilczek Nucl. Phys. **B479** (1996), 529.

[14] D. Ivanov, Phys. Rev. Lett. **86** (2001), 268.

[15] F. Wilczek, paper in preparation.

Frank Wilczek
Department of Physics
Massachusetts Institute of Technology
Cambridge, MA 02139, USA
e-mail: `wilczek@mit.edu`

The Spin
B. Duplantier, J.-M. Raimond and V. Rivasseau, Eds., 71–103

Poincaré Seminar 2007

Anyons and Lowest Landau Level Anyons

Stéphane Ouvry

Abstract. Intermediate statistics interpolating from Bose statistics to Fermi statistics are allowed in two dimensions. This is due to the particular topology of the two dimensional configuration space of identical particles, leading to non trivial braiding of particles around each other. One arrives at quantum many-body states with a multivalued phase factor, which encodes the anyonic nature of particle windings. Bosons and fermions appear as two limiting cases. Gauging away the phase leads to the so-called anyon model, where the charge of each particle interacts "à la Aharonov-Bohm" with the fluxes carried by the other particles. The multivaluedness of the wave function has been traded off for topological interactions between ordinary particles. An alternative Lagrangian formulation uses a topological Chern-Simons term in 2+1 dimensions. Taking into account the short distance repulsion between particles leads to an Hamiltonian well defined in perturbation theory, where all perturbative divergences have disappeared. Together with numerical and semi-classical studies, perturbation theory is a basic analytical tool at disposal to study the model, since finding the exact N-body spectrum seems out of reach (except in the 2-body case which is solvable, or for particular classes of N-body eigenstates which generalize some 2-body eigenstates). However, a simplification arises when the anyons are coupled to an external homogeneous magnetic field. In the case of a strong field, by projecting the system on its lowest Landau level (LLL, thus the LLL-anyon model), the anyon model becomes solvable, i.e., the classes of exact eigenstates alluded to above provide for a complete interpolation from the LLL-Bose spectrum to the LLL-Fermi spectrum. Being a solvable model allows for an explicit knowledge of the equation of state and of the mean occupation number in the LLL, which do interpolate from the Bose to the Fermi cases. It also provides for a combinatorial interpretation of LLL-anyon braiding statistics in terms of occupation of single particle states. The LLL-anyon model might also be relevant experimentally: a gas of electrons in a strong magnetic field is known to exhibit a quantized Hall conductance, leading to the integer and fractional quantum Hall effects. Haldane/exclusion statistics, introduced to describe FQHE edge excitations, is a priori different from anyon statistics, since it is not defined by braiding considerations, but rather by counting arguments in the space of available states. However, it has been shown to lead to the same kind of thermodynamics as the LLL-anyon thermodynamics (or, in other words, the

LLL-anyon model is a microscopic quantum mechanical realization of Haldane's statistics). The one dimensional Calogero model is also shown to have the same kind of thermodynamics as the LLL-anyons thermodynamics. This is not a coincidence: the LLL-anyon model and the Calogero model are intimately related, the latter being a particular limit of the former. Finally, on the purely combinatorial side, the minimal difference partition problem — partition of integers with minimal difference constraints on their parts — can also be mapped on an abstract exclusion statistics model with a constant one-body density of states, which is neither the LLL-anyon model nor the Calogero model.

1. Introduction

Quantum statistics, which is concerned with quantum many-body wavefunctions of identical particles, has a long history going back to Bose and Fermi. The concept of statistics originates at the classical level in the Gibbs paradox, which is solved by means of the indiscernibility postulate for identical particles. At the quantum level, the usual reasoning shows that only two types of statistics can exist, bosonic or fermionic. Indeed, since

- interchanging the positions of two identical particles can only amount to multiplying their 2-body wavefunction by a phase factor,
- a double exchange puts back the particles at their original position,
- and one usually insists on the univaluedness of the wavefunction,

this phase factor can be only 1 (boson) or -1 (fermion).

However, non trivial phase factor should be possible, since wavefunctions are anyway defined up to a phase. The configuration space of two, or more generally, N identical particles has to be defined cautiously [1]: denoting by C the configuration space of a single particle ($C = R^2$ for particles in the two-dimensional plane, $d = 2$), the configuration space of N particles should be of the type C^N/S_N, where $C \times C \times \ldots \times C = C^N$ and S_N is the permutation group for N identical particles. Division by S_N takes into account the identity of the particles which implies that one cannot distinguish between two configurations related by an operation of the permutation group. One should also subtract from C^N the diagonal of the configuration space D_N, i.e., any configurations where two or more particles coincide. The reason is, having in mind Fermi statistics, that the Pauli exclusion principle should be enforced in some way. A more precise argument is to have a valid classification of paths in the N-particle configuration space, which would be ambiguous if two or more particles coinciding at some time is allowed (since they are identical, did they cross each other, or did they scatter off each other ?). It follows that the configuration space of N identical particles should be

$$\tilde{C}_N = \frac{C^N - D_N}{S_N} . \tag{1}$$

Note that on this configuration space, a fermionic wavefunction is multivalued (two values 1 and -1), so there is no reason not to allow more general multivaluedness. Here come some topological arguments, which allow to distinguish between $d = 2$ and $d > 2$, and, as we will see later, which can be related to spin considerations. In 2 dimensions, C^N is multiply connected and its topology is non trivial: it is not possible to shrink a path of a particle encircling another particle, due to the topological obstruction materialized by the latter. It follows that $\tilde{C}_N$ is multiply connected. This is not the case in dimensions higher than 2, where C^N is simply connected, meaning that all paths made by a particle can be continuously deformed into each other, i.e., one cannot distinguish the interior from the exterior of a closed path of a particle around other particles.

These arguments imply that the equivalent classes of paths (first homotopy group) in $\tilde{C}_N$ are, when $d = 2$, in one-to-one correspondence with the elements of the braid group

$$\Pi_1(\tilde{C}_N) = B_N \tag{2}$$

whereas, when $d > 2$, they are in one-to-one correspondence with the elements of the permutation group

$$\Pi_1(\tilde{C}_N) = S_N. \tag{3}$$

The braid group generators σ_i interchange the position of particle i with particle $i+1$. This operation can be made in an anti-clockwise manner (σ_i) or a clockwise manner (σ_i^{-1}). Each braiding of N particles consists of a sequence of interchanges of pairs of neighboring particles via the σ_i's and the σ_i^{-1}'s, with $i = 1, 2, ..., N-1$. The braid group relations list the equivalent braiding, i.e., braiding that can be continuously deformed one into the other without encountering a topological obstruction

$$\sigma_i\sigma_{i+1}\sigma_i = \sigma_{i+1}\sigma_i\sigma_{i+1}; \qquad \sigma_i\sigma_j = \sigma_j\sigma_i \quad \text{when} \quad |i-j| > 2. \tag{4}$$

Saying that $d = 2$ is different from $d > 2$ is nothing but recognizing that $\sigma_i \neq \sigma_i^{-1}$ when $d = 2$, whereas $\sigma_i = \sigma_i^{-1}$ when $d > 2$ (σ_i can be continuously deformed into σ_i^{-1} when particles i and $i+1$ are not stuck in a plane). It follows that when $d > 2$, the braid generators σ_i's defined by (4) with the additional constraint $\sigma_i = \sigma_i^{-1}$ are the permutation group generators.

Note also that the $d = 2$ paradigm, $\sigma_i \neq \sigma_i^{-1}$, hints at an orientation of the plane, a hallmark of the presence of some sort of magnetic field. This point will become apparent in the Aharonov-Bohm formulation of the anyon model.

The fact that C^N is multiply connected when $d = 2$ and not when $d > 2$ can also be related to the rotation group $O(d)$, and thus to some spin-statistics considerations [2]. When $d > 2$, the rotation group is doubly connected, $[\Pi_1(O(d)) = Z_2]$, its universal covering, for example when $d = 3$, is $SU(2)$, which allows for either integer or half integer angular momentum states, that is to say either single valued or double valued representations of the rotation group. On the other hand, when $d = 2$, the rotation group is Abelian and infinitely connected $[\Pi_1(O(2)) = Z]$, its universal covering is the real line, that is to say arbitrary angular momenta are

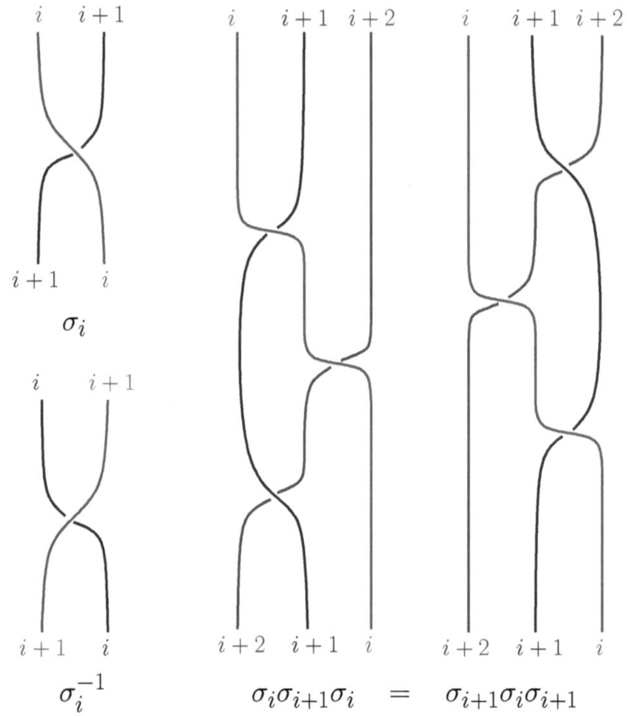

FIGURE 1. The braid group generators and their defining relations.

possible, and therefore multivalued representations. One can see here a hint about the spin-statistics connection, where statistics and spin are trivial (Bose-Fermi statistics, integer-half integer spin) when $d > 2$, and not when $d = 2$.

Let us consider the simple one-dimensional irreducible representation of the braid group, which amounts to a common phase factor $\exp(-i\pi\alpha)$ for each generator σ (and thus $\exp(i\pi\alpha)$ to σ^{-1}). It means that a non trivial phase has been associated with the winding of particle i around particle $i+1$. Higher dimensional representations (quantum vector states) are possible -one speaks of non Abelian anyons, in that case not only a non trivial phase materializes during a winding, but also the direction of the vector state in the Hilbert space is affected- but they will not be discussed here (even though they might play a role in the discussion of certain FQHE fractions [3], and, in a quite different perspective, in the definition of topologically protected fault-tolerant quantum computers [4]).

Clearly, when $d > 2$, $\sigma_i = \sigma_i^{-1}$ implies $\alpha = 0$ or $\alpha = 1$, i.e., Bose or Fermi statistics (an interchange leaves the wavefunction unchanged or affected by a minus sign).

From now on let us concentrate on $d = 2$ and denote the free many-body wavefunction of N identical particles by $\psi'(\vec{r}_1, \vec{r}_2, ..., \vec{r}_N)$. Indeed, statistics should be defined for free particles with Hamiltonian

$$H'_N = \sum_{i=1}^{N} \frac{\vec{p_i}^2}{2m} \tag{5}$$

and special boundary conditions on the wavefunction, as in the Bose case (symmetric boundary condition) and the Fermi case (antisymmetric boundary condition). As already said, $\psi'(\vec{r}_1, \vec{r}_2, ..., \vec{r}_N)$ is affected by a phase $\exp(-i\pi\alpha)$ when particles i and $i+1$ are interchanged: one can encode this non trivial exchange property by defining

$$\psi'(\vec{r}_1, \vec{r}_2, ..., \vec{r}_N) = \exp(-i\alpha \sum_{i<j} \theta_{ij})\psi(\vec{r}_1, \vec{r}_2, ..., \vec{r}_N) \tag{6}$$

where $\psi(\vec{r}_1, \vec{r}_2, ..., \vec{r}_N)$ is a regular wavefunction, say bosonic by convention, and θ_{ij} is the angle between the vector $\vec{r}_j - \vec{r}_i \equiv \vec{r}_{ij}$ and a fixed direction in the plane. Indeed interchanging i with j amounts to $\theta_{ij} \to \theta_{ij} \pm \pi$, which altogether with the bosonic symmetry of $\psi(\vec{r}_1, \vec{r}_2, ..., \vec{r}_N)$, leads to

$$\psi'(\vec{r}_1, \vec{r}_2, ...\vec{r}_j, ...\vec{r}_i, ..., \vec{r}_N) = \exp(\mp i\pi\alpha)\psi'(\vec{r}_1, \vec{r}_2, ...\vec{r}_i, ...\vec{r}_j, ..., \vec{r}_N) \tag{7}$$

By the above bosonic convention for $\psi(\vec{r}_1, \vec{r}_2, ..., \vec{r}_N)$, the statistical parameter α even (odd) integer corresponds to Bose (Fermi) statistics. It is defined modulo 2, since two quanta of flux can always be gauged away by a regular gauge transformation while preserving the symmetry of the wavefunctions in the Bose or Fermi systems. Indeed, (6) can be interpreted as a gauge transformation. Let us compute the resulting Hamiltonian H_N acting on $\psi(\vec{r}_1, \vec{r}_2, ..., \vec{r}_N)$

$$H_N = \sum_{i=1}^{N} \frac{1}{2m}(\vec{p}_i - \vec{A}(\vec{r}_i))^2 \tag{8}$$

where

$$\vec{A}(\vec{r}_i) = \alpha\vec{\partial}_i \left(\sum_{k<l} \theta_{kl} \right) = \alpha \sum_{j,j\neq i} \frac{\vec{k} \wedge \vec{r}_{ij}}{r_{ij}^2} \tag{9}$$

is the statistical potential vector associated with the multivalued phase (the gauge parameter). The free multivalued wavefunction has been traded off for a regular bosonic wavefunction with topological singular magnetic interactions. The statistical potential vector (9) can be viewed as the Aharonov-Bohm (A-B) potential vector that particle i carrying a charge e would feel due to the flux tube ϕ carried by the other particles, with e and ϕ related to the statistical parameter α by $\alpha = e\phi/(2\pi) = \phi/\phi_0$ ($\phi_0 = 2\pi/e$ is the flux quantum in units $\hbar = 1$). The resulting composite charge-flux picture is known under the name of anyon model [5] since it describes particles with "any" (any-on) statistics.

Computing the field strength one obtains

$$\frac{\alpha}{e}\vec{\partial}_i \wedge \sum_{j,j\neq i} \frac{\vec{k} \wedge \vec{r}_{ij}}{r_{ij}^2} = \frac{2\pi\alpha}{e} \sum_{j,j\neq i} \delta(\vec{r}_{ij}) \tag{10}$$

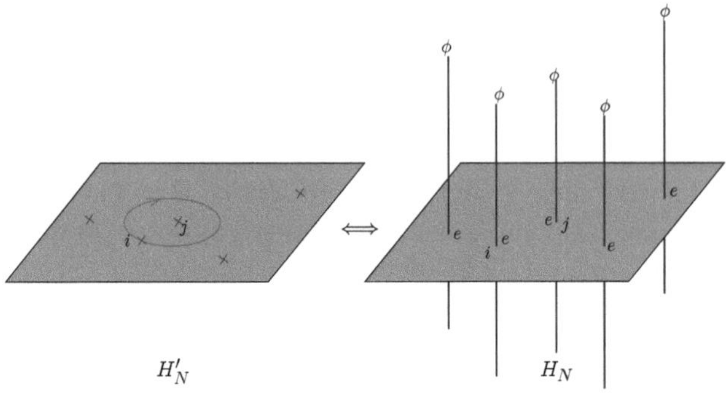

FIGURE 2. The two equivalent formulations of anyon statistics in terms, on the left, of a punctured plane and, on the right, of usual bosonic particles interacting via topological A-B interactions. The loop of particle i around particle j cannot be continuously deformed to nothing due to the topological obstruction materialized by the puncture at the location of particle j.

meaning that each particle carries an infinite singular magnetic field with flux $\phi = 2\pi\alpha/e$. The gauge transformation is singular since it does not preserve the field strength (which vanishes in the multivalued gauge and is singular in the regular gauge). This is due to the singular behavior of the gauge parameter θ_{ij} when particle i come close to particle j, thus the singular Dirac $\delta(\vec{r}_{ij})$ function in the field strength.

It is not surprising that topological A-B interactions are at the heart of quantum statistics. In its original form, the A-B effect [6] consists in the phase shift in electron interference due to the electromagnetic field, determined by the phase factor $\exp[(ie/\hbar c)\int_\gamma A_\mu dx^\mu]$ along a closed curve γ passing through the beam along which the field strength vanishes. This effect[1] is counter-intuitive to the usual understanding that the influence of a classical electromagnetic field on a charged particle can only occur through the local action of the field strength. In the context of quantum statistics, it means that non trivial statistics arise through topological "infinite"-distance interactions where no classical forces are present, as it should and as it is the case for Bose and Fermi statistics. Finally, singular magnetic fields give an orientation to the plane, which, as already said, shows up in $\sigma_i \neq \sigma_i^{-1}$.

All this can be equivalently restated in a Lagrange formulation which describes again the system in topological terms, i.e., free particles minimally coupled to a potential vector whose dynamics is not Coulomb-like (Maxwell Lagrangian)

[1]The effect was first experimentally confirmed by R. G. Chambers [7], then by A. Tonomura [8].

but rather Chern-Simons [9]

$$L_N = \sum_{i=1}^{N} \left(\frac{1}{2} m \vec{v}_i^2 + e \left(\vec{A}(\vec{r}_i)\vec{v}_i - A^0(\vec{r}_i) \right) \right) - \frac{\kappa}{2} \epsilon_{\mu\nu\rho} \int d^2\vec{r} A^\mu \partial^\nu A^\rho \tag{11}$$

with $\epsilon_{\mu\nu\rho}$ the completely antisymmetric tensor (the metric is $(+,-,-), x^\mu = (t,\vec{r}) = (t,x,y), A_\mu = (A_o, A_x, A_y), \epsilon_{012} = \epsilon^{012} = +1$). Solving the Euler-Lagrange equations, in particular

$$\partial^\sigma \frac{\delta L_N}{\delta(\partial^\sigma A_0)} = \frac{\delta L_N}{\delta A^0} \to \kappa \vec{\partial} \wedge \vec{A}(\vec{r}) = e \sum_{j=1}^{N} \delta(\vec{r} - \vec{r}_j) \tag{12}$$

leads to a magnetic field proportional to the density of particles in accordance with (10). Solving this last equation for $\vec{A}(\vec{r})$ in the Coulomb gauge gives

$$\vec{A}(\vec{r}) = \frac{e}{2\pi\kappa} \sum_{j=1}^{N} \frac{\vec{k} \wedge (\vec{r} - \vec{r}_j)}{(\vec{r} - \vec{r}_j)^2} \tag{13}$$

in accordance with the A-B potential vector (9). Here again there is no Lorentz force, the potential vector is a pure gauge, the Chern-Simons term is metric independent, and the field strength is directly related to the matter current.

Coming back to the Hamiltonian formulation (8), one might ask how the exclusion of the diagonal of the configuration space materializes in the Hamiltonian formulation. One way to look at it is perturbation theory [10, 11]. Let us simplify the problem by considering the standard A-B problem, i.e., a charged particle in the plane coupled to a flux tube at the origin with the Hamiltonian

$$H = \frac{1}{2m} \left(\vec{p} - \alpha \frac{\vec{k} \wedge \vec{r}}{r^2} \right)^2 . \tag{14}$$

Let us see what happens close to Bose statistics when $\alpha \simeq 0$ (by periodicity α can always been chosen in $[-1/2, +1/2]$, an interval of length 1, since in the one-body case one quantum of flux can always be gauged away via a regular gauge transformation). The A-B spectrum [6] is given by the Bessel functions

$$\psi(\vec{r}) = e^{il\theta} J_{|l-\alpha|}(kr) \qquad E = \frac{k^2}{2m} \tag{15}$$

with wavefunctions vanishing close to the origin $r \to 0$ as $J_{|l-\alpha|}(kr) \simeq r^{|l-\alpha|}$. When the angular momentum $l \neq 0$, this is the only possible locally square-integrable function. However, when $l = 0$, one could have as well $J_{-|\alpha|}(kr)$ as a solution, since it is still locally square-integrable even though it diverges at the origin as $r^{-|\alpha|}$. In principle, the general solution in the $l = 0$ sector should be a linear combination of $J_{|\alpha|}(kr)$ and $J_{-|\alpha|}(kr)$, introducing an additional scale in the coefficient of the linear combination [12]. Restricting the space of solutions as in (15), i.e., wavefunctions vanishing at the origin, means that a short-range repulsive prescription has been imposed on the behavior of the wavefunctions when the particle comes close to the flux tube. One can give a more precise formulation of this fact by trying to compute in perturbation theory the spectrum (15). Expanding the square in the Hamiltonian (14), one finds that the α^2/r^2 term, which is as singular as the kinetic term, is divergent at second order in perturbation theory in the $l=0$ sector. It follows that perturbation theory is not well defined in the problem as defined by the Hamiltonian (14). A renormalization has to be implemented: one realizes that by adding the counterterm $\pi|\alpha|\delta(\vec{r})$ to (14), i.e., by considering

$$H = \frac{1}{2m}(\vec{p} - \alpha\frac{\vec{k}\wedge\vec{r}}{r^2})^2 + \frac{2\pi|\alpha|}{m}\delta(\vec{r}) \tag{16}$$

the perturbative divergences due to the α^2/r^2 term are exactly cancelled by those arising from the $\pi|\alpha|\delta(\vec{r})$ term at all orders in perturbation theory, giving back the spectrum (15). Physically, this repulsive δ contact term means that the particle is prevented from penetrating the core of the flux tube where the field strength is infinite, thus the (at least) $r^{|\alpha|}$ behavior when $r \to 0$. Note that this has been achieved without introducing any additional scale in the problem.

Clearly, in the N-body A-B anyon formulation of the model, the corresponding renormalized Hamiltonian should read

$$H_N = \sum_{i=1}^{N} \frac{1}{2m}\left(\vec{p}_i - \alpha\sum_{j\neq i}\frac{\vec{k}\wedge\vec{r}_{ij}}{r_{ij}^2}\right)^2 + \frac{2\pi|\alpha|}{m}\sum_{i\neq j}\delta(\vec{r}_{ij}) \tag{17}$$

realizing the quantum mechanical exclusion of the diagonal of the configuration space in terms of contact repulsive interaction between particles. Note that the term $\pi|\alpha|\sum_{i\neq j}\delta(\vec{r}_{ij})$ in (17) can also be viewed [11] as the Pauli spin coupling of the spin of the particles to the singular magnetic field (10) associated to the flux tubes.

The anyon model defined in (17) is properly defined as far as short-distance considerations are concerned. It is the interacting formulation for regular wavefunctions of the free particles formulation for multivalued wavefunctions. Both Hamiltonians H_N and H'_N are equivalent, the former being more familiar in terms of usual quantum mechanics, the latter more relevant to study braiding and winding properties.

The anyon model has been the subject of numerous studies in the eighties and the nineties [13], some of them analytical, starting with the 2-body case which is solvable since its relative 2-body problem is the usual A-B problem (14) with an even (Bose) angular momentum l. The exact solution [1, 5] for the relative 2-body problem is given by (15), l being an even integer, therefore when α is odd, $l-\alpha$ is odd, corresponding to Fermi statistics (the periodicity $\alpha \to \alpha \pm 2p$ is manifest in the shift $l \to l-\alpha$). These studies were followed by the 3-body [14] and then the N-body problem [15]. Statistical mechanics was also considered (second virial coefficient [16, 17], third virial coefficient [18]). However, it soon became apparent that a complete N-body spectrum was out of reach, to the exception of particular classes of exact eigenstates generalizing the 2-body eigenstates. Numerical [19] as well as semi-classical [20] studies were performed giving indications on the low energy N-body spectrum. A systematic study of the model was achieved at first [21] and at second order [22] in perturbation theory (at second order the complexity of the model shows up clearly). Numerical studies [23], taking some input from the perturbative results, were performed for the 3rd and 4th virial coefficients. Last but not least, on the experimental side, Laughlin quasiparticles [24] were put forward as the elementary excitations of highly-correlated fractional quantum Hall electron fluids [25]. They were supposed to carry a fractional charge and to obey anyon statistics [26], a fact confirmed by Berry phase calculations, at least for quasiholes [27] (for quasiparticles the situation is less clear). The quasiparticles can propagate quantum-coherently in chiral edge channels, and constructively or destructively interfere. Unlike electrons, the interference condition for Laughlin quasiparticles has a non-vanishing statistical contribution which might be observed experimentally [28].

Some kind of simplification had to be made to render the model more tractable, and possibly solvable, at least in a certain sector. One realized that this was the case if one considered, in addition to the singular statistical magnetic field, an external homogeneous magnetic field perpendicular to the plane, to which the charge of the anyons couple. In the case of a strong magnetic field, by projecting the system of anyons coupled to the magnetic field in its LLL, the model becomes solvable meaning that one can find a class of N-body eigenstates which interpolates continuously from the LLL-Bose to the LLL-Fermi eigenstates basis: this is the LLL-anyon model [29].

2. The LLL-anyon model

From now on, let us set the mass of the particles $m = 1$ and choose the statistical parameter $\alpha \in [-2, 0]$. It is understood that all the results below are obtained for α in this interval, but they can be periodically continued to the whole real axis. Before introducing an external magnetic field, let us come back to the anyon Hamiltonian (17) and take advantage of wavefunctions vanishing at least as $r_{ij}^{-\alpha}$ when $r_{ij} \to 0$ (exclusion of the diagonal of the configuration space in the quantum mechanical formulation) by encoding this short distance behavior in the N-body bosonic wavefunction [10]

$$\psi(\vec{r}_1, \vec{r}_2, ..., \vec{r}_N) = \prod_{i<j} r_{ij}^{-\alpha} \tilde{\psi}(\vec{r}_1, \vec{r}_2, ..., \vec{r}_N). \tag{18}$$

$\tilde{\psi}(\vec{r}_1, \vec{r}_2, ..., \vec{r}_N)$ is regular but does not have to vanish at coinciding points. From H_N in (17) one can compute the new Hamiltonian $\tilde{H}_N$ acting on $\tilde{\psi}(\vec{r}_1, \vec{r}_2, ..., \vec{r}_N)$. Since H_N is itself obtained from the free Hamiltonian H'_N in (5) via the singular gauge transformation (6), it is more transparent to start directly from the free formulation. In complex notation (the free Hamiltonian is $H'_N = -2\sum_{i=1}^{N} \partial\bar{\partial}_i$)

the wavefunction redefinitions (6) and (18) combined take the simple form

$$\psi'(z_1, z_2, ..., z_N; \bar{z}_1, \bar{z}_2, ..., \bar{z}_N) = \prod_{i<j} z_{ij}^{-\alpha} \tilde{\psi}(z_1, z_2, ..., z_N; \bar{z}_1, \bar{z}_2, ..., \bar{z}_N). \qquad (19)$$

The Jastrow-like prefactor $\prod_{i<j} z_{ij}^{-\alpha}$ in (19) encodes in the wavefunction the essence of anyon statistics: topological braiding phase and short-distance contact exclusion behavior. It is immediate that $\tilde{H}_N$ rewrites as

$$\tilde{H}_N = -2\sum_{i=1}^{N} \partial_i \bar{\partial}_i + 2\alpha \sum_{i<j} \frac{1}{z_i - z_j}(\bar{\partial}_i - \bar{\partial}_j). \qquad (20)$$

It is a non-Hermitian Hamiltonian (the transformation (19) is non-unitary), but it has a simple form, linear in α and well defined in perturbation theory (it is perturbatively divergence free). Any analytic wavefunction of the z_i's is a N-body eigenstate of $\tilde{H}_N$, and therefore of the N-anyon Hamiltonian (17) taking into account (18). Analytical eigenstates are known to live in the LLL of a magnetic field, if such a field were present. Let us couple the electric charge of each anyon to an external magnetic field B perpendicular to the plane such that by convention $eB > 0$ and let us denote by $\omega_c = eB/2$ half its cyclotron frequency. One now starts from the Landau Hamiltonian

$$H'_N = -2\sum_i \left(\partial_i - \frac{\omega_c}{2}\bar{z}_i\right)\left(\bar{\partial}_i + \frac{\omega_c}{2} z_i\right). \qquad (21)$$

In a magnetic field, the 1-body eigenstates have a long-distance Landau exponential behavior $\exp(-\frac{1}{2}\omega_c z_i \bar{z}_i)$. Let us also encode this behavior in the wavefunction redefinition (19) so that it becomes

$$\psi'(z_1, z_2, ..., z_N; \bar{z}_1, \bar{z}_2, ..., \bar{z}_N)$$
$$= \prod_{i<j} z_{ij}^{-\alpha} \exp\left(-\frac{1}{2}\omega_c \sum_{i=1}^{N} z_i \bar{z}_i\right) \tilde{\psi}(z_1, z_2, ..., z_N; \bar{z}_1, \bar{z}_2, ..., \bar{z}_N). \qquad (22)$$

One obtains

$$\tilde{H}_N = -2\sum_{i=1}^{N}(\partial_i \bar{\partial}_i - \omega_c \bar{z}_i \bar{\partial}_i) + 2\alpha \sum_{i<j} \frac{1}{z_i - z_j}(\bar{\partial}_i - \bar{\partial}_j) + N\omega_c \qquad (23)$$

where the trivial constant energy shift from the Pauli coupling to the magnetic field has been ignored. As announced, $\tilde{H}_N$ acts trivially on N-body eigenstates made of symmetrized products of analytic 1-body LLL eigenstates

$$\sqrt{\frac{\omega_c^{l_i+1}}{\pi l_i!}} z_i^{l_i}; \quad l_i \geq 0; \quad E = \omega_c \qquad (24)$$

(in (24) the Landau exponential term is missing since it has already been taken into account in (22)). So, up to an overall normalization,

$$\tilde{\psi}(z_1, z_2, ..., z_N; \bar{z}_1, \bar{z}_2, ..., \bar{z}_N) = \mathrm{Sym} \prod_{i=1}^{N} z_i^{l_i}; \quad 0 \le l_1 \le l_2 \le ... \le l_N \tag{25}$$

is an eigenstate with a degenerate N-body energy, $E_N = N\omega_c$, a mere reflection of the fact that there are N particles in the LLL. From (22) and (25) one finally gets

$$\psi'(z_1, z_2, ..., z_N; \bar{z}_1, \bar{z}_2, ..., \bar{z}_N) = \prod_{i<j} z_{ij}^{-\alpha} \exp\left(-\frac{1}{2}\omega_c \sum_{i=1}^{N} z_i \bar{z}_i\right) \mathrm{Sym} \prod_{i=1}^{N} z_i^{l_i}; \quad 0 \le l_1 \le l_2 \le ... \le l_N. \tag{26}$$

The basis (26) continuously interpolates when $\alpha = 0 \to -1$ from the complete LLL-Bose N-body basis to the complete LLL-Fermi N-body basis. Indeed, when $\alpha = -1$,

$$\psi'(z_1, z_2, ..., z_N; \bar{z}_1, \bar{z}_2, ..., \bar{z}_N) = \exp\left(-\frac{1}{2}\omega_c \sum_{i=1}^{N} z_i \bar{z}_i\right) \prod_{i<j} z_{ij} \mathrm{Sym} \prod_{i=1}^{N} z_i^{l_i}; \quad 0 \le l_1 \le l_2 \le ... \le l_N \tag{27}$$

is equivalent to

$$\psi'(z_1, z_2, ..., z_N; \bar{z}_1, \bar{z}_2, ..., \bar{z}_N) = \exp\left(-\frac{1}{2}\omega_c \sum_{i=1}^{N} z_i \bar{z}_i\right) \mathrm{Antisym} \prod_{i=1}^{N} z_i^{l'_i}; \quad 0 < l'_1 < l'_2 < ... < l'_N \tag{28}$$

i.e., the LLL fermionic basis. One has therefore obtained a complete LLL-Bose $\to$ LLL-Fermi interpolating basis which allows, in principle, for a complete knowledge of the LLL-anyon system with statistics intermediate between Bose and Fermi statistics.

One could ask about going beyond the Fermi point $\alpha = -1$ up to the Bose point $\alpha = -2$. This question is related to the validity of the LLL projection, since ignoring higher Landau levels amounts to assuming that excited non LLL states above the N-body LLL ground state have a non vanishing gap. Considerations around the Fermi point, as well as numerical and semiclassical analysis, support [29] this scheme as long as α does not come close to -2. However, when $\alpha \to -2$, known linear as well as unknown nonlinear non LLL eigenstates do join the LLL ground state [31]. Said differently, the LLL-anyon basis (26) does not constitute a complete LLL-Bose basis when $\alpha \to -2$, i.e., some N-body LLL bosonic quantum numbers are missing at this point. We will come back to this issue later.

One has not seen yet any α dependence in the N-body energy, a situation already encountered in the 1-body A-B problem, where the free continuous energy spectrum (15) is α-independent. This is due to the fact that a magnetic field does

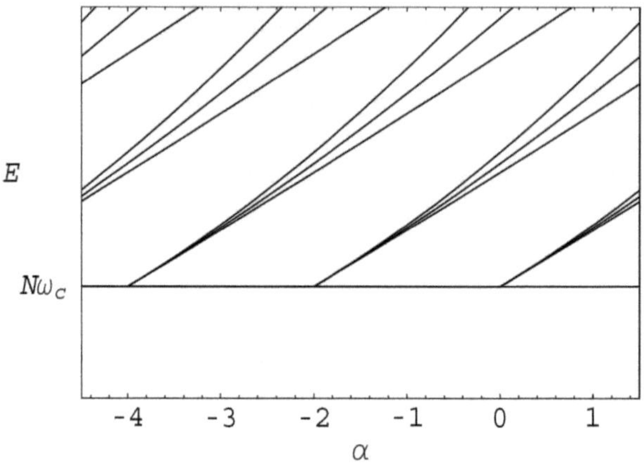

FIGURE 3. Linear and non linear non LLL eigenstates merge in the LLL ground state at the bosonic values of α.

not confine particles: classical orbits are circular cyclotron orbits, but their centers, due to translation invariance, are located anywhere in the plane. Translation invariance in turn gives, in quantum mechanics, a Landau spectrum which is l_i independent, and therefore infinitely degenerate[2]. The degeneracy factor scales as the infinite surface V of the 2d sample: it is the flux of the magnetic field counted in units of the flux quantum $\phi_0 = 2\pi/e$ (in units $\hbar = 1$)

$$N_L = \frac{VB}{\phi_0}. \tag{29}$$

Statistical interactions being topological interactions, one does not expect, in the infinite plane limit, any effect on the N-body energies. To see such an effect, one has to introduce a long-distance confinement, like putting the particles in a box. Let us rather introduce [30] a more convenient harmonic well confinement where the particles are trapped, so that the Landau Hamiltonian (21) becomes

$$H'_N = -2\sum_{i=1}^{N}\left(\partial_i - \frac{eB}{4}\bar{z}_i\right)\left(\bar{\partial}_i + \frac{eB}{4}z_i\right) + \frac{1}{2}\omega^2\sum_{i=1}^{N} z_i\bar{z}_i. \tag{30}$$

The virtue of the harmonic confinement is to lift the degeneracy with respect to the angular momentum l_i of the 1-body Landau eigenstates: the harmonic LLL

[2]From this point of view one can argue that the Landau spectrum is continuous, albeit being made of discrete Landau levels, due to the infinite degeneracy on each level.

spectrum[3] becomes

$$\sqrt{\frac{\omega_t^{l_i+1}}{\pi l_i!}} z_i^{l_i} \exp\left(-\frac{1}{2}\omega_t z_i \bar{z}_i\right); \quad l_i \geq 0; \quad E = (\omega_t - \omega_c)(l_i + 1) + \omega_c \tag{31}$$

where $\omega_t = \sqrt{\omega^2 + \omega_c^2}$. Each harmonic LLL level in (31) has now a finite degeneracy, with an eigenstate still analytic in z_i, up to the long-distance harmonic Landau combined exponential behavior. Let us take into account this exponential behavior in the redefinition of the free N-body wavefunction so that (22) now becomes

$$\psi'(z_1, z_2, ..., z_N; \bar{z}_1, \bar{z}_2, ..., \bar{z}_N)$$
$$= \prod_{i<j} z_{ij}^{-\alpha} \exp\left(-\frac{1}{2}\omega_t \sum_{i=1}^{N} z_i \bar{z}_i\right) \tilde{\psi}(z_1, z_2, ..., z_N; \bar{z}_1, \bar{z}_2, ..., \bar{z}_N). \tag{32}$$

Starting from the Hamiltonian (30) one obtains [32, 33]

$$\tilde{H}_N = -2\sum_{i=1}^{N} \left(\partial_i \bar{\partial}_i - \frac{\omega_t + \omega_c}{2} \bar{z}_i \bar{\partial}_i - \frac{\omega_t - \omega_c}{2} z_i \partial_i\right)$$
$$+ 2\alpha \sum_{i<j} \left[\frac{1}{z_i - z_j}(\bar{\partial}_i - \bar{\partial}_j) - \frac{\omega_t - \omega_c}{2}\right] + N\omega_c. \tag{33}$$

Again let us act on N-body eigenstates made, in analogy with (25), of symmetrized products of the 1-body harmonic LLL eigenstates (31)

$$\tilde{\psi}(z_1, z_2, ..., z_N; \bar{z}_1, \bar{z}_2, ..., \bar{z}_N) = \mathrm{Sym} \prod_{i=1}^{N} z_i^{l_i}; \quad 0 \leq l_1 \leq l_2 \leq ... \leq l_N. \tag{34}$$

Acting on this basis, the Hamiltonian (33) rewrites as

$$\tilde{H}_N = (\omega_t - \omega_c)\left[\sum_{i=1}^{N} z_i \partial_i - \alpha\frac{N(N-1)}{2} + N\right] + N\omega_c, \tag{35}$$

so that the N-anyon energy spectrum is

$$E_N = (\omega_t - \omega_c)\left[\sum_{i=1}^{N} l_i - \alpha\frac{N(N-1)}{2} + N\right] + N\omega_c. \tag{36}$$

The N-anyon spectrum (36) is a sum of 1-body harmonic LLL spectra shifted by the 2-body statistical term $-(\omega_t - \omega_c)\alpha N(N-1)/2$. The effect of the harmonic well has been not only to lift the degeneracy with respect to the l_i's, but also

---

[3]The complete 2d harmonic Landau spectrum is, with the convention $eB > 0$,

$$\omega_t(2n_i + l_i + 1) - l_i\omega_c; \quad n_i \geq 0, \quad l_i \in Z.$$

The LLL quantum numbers are $n_i = 0$ and $l_i \geq 0$.

to make the energy dependence on α explicit. When computing thermodynamical quantities like the equation of state, the harmonic well regulator will also be needed to compute finite quantities in a finite "harmonic" box, and then take the thermodynamic limit, by letting $\omega \to 0$ in an appropriate way.

The resulting eigenstates from (32)

$$\psi'(z_1, z_2, ..., z_N; \bar{z}_1, \bar{z}_2, ..., \bar{z}_N) = \prod_{i<j} z_{ij}^{-\alpha} \exp\left(-\frac{1}{2}\omega_t \sum_{i=1}^{N} z_i \bar{z}_i\right) \text{Sym} \prod_{i=1}^{N} z_i^{l_i};$$
$$0 \leq l_1 \leq l_2 \leq ... \leq l_N \quad (37)$$

are called "linear states" since their energy (36) varies linearly with α. As already stressed, they constitute a set of exact N-body eigenstates which is only a small part of the complete N-body spectrum, which remains mostly unknown. However, what makes, in the LLL context, these linear states particularly interesting is that they continuously interpolate when $\alpha = 0 \to -1$ from the complete harmonic LLL-Bose basis to the complete harmonic LLL-Fermi basis.

Before turning to LLL-anyon thermodynamics, let us reconsider the physical charge-flux composite interpretation of the anyon model, where the charges are now coupled to an external magnetic field. A given particle, say the Nth, sees a "positive" ($eB > 0$) magnetic field perpendicular to the plane, and $N-1$ "negative" ($e\phi = 2\pi\alpha < 0$, $\alpha \in [-1, 0]$) point vortices piercing the plane at the positions of the other particles. This is a screening regime: in the large N limit where a mean field picture is expected to be valid, the more α is close to the fermionic point $\alpha = -1$, the more the external magnetic field is screened by the mean magnetic field associated with the vortices. In terms of the total (external + mean) magnetic field $\langle B \rangle$ that the Nth particle sees, or rather in terms of its flux $V\langle B \rangle$, or, when counted in units of the flux quantum, in terms of the Landau degeneracy $\langle N_L \rangle$, one has

$$V\langle B \rangle/\phi_0 = (VB)/\phi_0 + (N-1)\phi/\phi_0 \quad \text{i.e.} \quad \langle N_L \rangle = N_L + (N-1)\alpha. \qquad (38)$$

Moving away from the Bose point, i.e., $\alpha \leq 0$, as N increases the number $\langle N_L \rangle$ of 1-body quantum states available for the Nth particle in the LLL of $\langle B \rangle$ decreases. This sounds reasonable, bearing in mind that a fermion occupies a quantum state to the exclusion of others (Pauli exclusion), whereas bosons can condense (Bose condensation). Introducing the LLL filling factor

$$\nu = \frac{N}{N_L} \qquad (39)$$

one deduces from (38) a maximal critical filling [29] for which the screening is total, $\langle N_L \rangle = 0$

$$\nu = -\frac{1}{\alpha}. \qquad (40)$$

This is nothing but recognizing once more that bosons ($\alpha = 0$) can infinitely fill a quantum state ($\nu = \infty$), whereas fermions ($\alpha = -1$) are at most one per quantum state ($\nu = 1$). In between, one finds that there are at most $-1/\alpha$ anyons per quantum state.

Interestingly enough, Haldane/exclusion statistics definition[4] happens to coincide with (38): for a gas of particles obeying Haldane/exclusion statistics [34] with statistical parameter $g \in [0,1]$, given N_L degenerate energy levels and $N-1$ particles already populating the levels, the number d_N of quantum states still available for the Nth particle is given by (38) where $-\alpha$ is replaced by g

$$d_N = N_L - (N-1)g. \tag{41}$$

On the one hand, Haldane's definition (41) stems from an arbitrary combinatorial point of view, inspired by the Bose and Fermi counting of states. On the other hand, in the LLL-anyon model, (38) is obtained from a somehow ad-hoc mean field ansatz. We will come back to these issues in the next section.

3. LLL-anyon thermodynamics

Let us rewrite the N-body energy (36) as [37]

$$E_N = \sum_{i=1}^{N} (\epsilon_0 + l_i \tilde{\omega}) - \alpha \frac{N(N-1)}{2} \tilde{\omega}; \quad 0 \le l_1 \le l_2 \le \ldots \le l_N \tag{42}$$

with $\tilde{\omega} = (\omega_t - \omega_c)$ and $\epsilon_0 = \omega_c$. Introducing the fugacity z and the inverse temperature β, one wants to compute the thermodynamic potential

$$\ln Z(\beta, z) = \ln \left(\sum_{N=0}^{\infty} z^N Z_N \right); \quad Z_0 = 1 \tag{43}$$

where $Z(\beta, z)$ is the grand partition function defined in terms of the N-body partition functions $Z_N = \mathrm{Tr}\exp(-\beta H'_N) = \mathrm{Tr}\exp(-\beta H_N) = \mathrm{Tr}\exp(-\beta \tilde{H}_N)$. The thermodynamic potential rewrites as $\ln Z(\beta, z) = \sum_{n=1}^{\infty} b_n z^n$ where, at order z^n, the cluster coefficient b_n only requires the knowledge of the Z_i's, with $i \le n$. One is interested in evaluating the thermodynamic potential in the thermodynamical limit, i.e., ω is small, which means, here, that the dimensionless quantity $\beta\omega$ is small. The N-body spectrum, as given in (42), allows to compute, at leading order in $\beta\omega \to 0$, the Z_i's for $i \le n$, and thus the b_n's

$$b_n = \frac{1}{\beta\tilde{\omega}} \frac{e^{-n\beta\epsilon_0}}{n^2} \prod_{k=1}^{n-1} \frac{k + n\alpha}{k}; \quad b_1 = \frac{1}{\beta\tilde{\omega}} e^{-\beta\omega_c}. \tag{44}$$

One has still to give a meaning, in the thermodynamic limit $\beta\omega = 0$, to the scaling factor $1/(\beta\tilde{\omega})$ in (44). To this purpose, one temporarily switches off the anyonic interaction and the external magnetic field, and considers a quantum gas of non interacting harmonic oscillators per se. One asks, when $\beta\omega \to 0$, for its cluster

[4] This is Haldane's statistics for one particle species. It can be generalized to the multispecies case.

coefficients to yield the infinite box (plane wave) cluster coefficients. At order n in the cluster expansion, in d dimensions, one obtains [10]

$$\lim_{\beta\omega\to 0}\left(\frac{1}{n(\beta\omega)^2}\right)^{\frac{d}{2}} = \frac{V}{\lambda^d} \tag{45}$$

where $\lambda = \sqrt{2\pi\beta}$ is the thermal wavelength and V is the d-dimensional infinite volume (in $d = 2$ dimensions, V is, as defined above, the infinite area of the 2d sample). Using the thermodynamic limit prescription (45), the cluster coefficient (44) rewrites, in the thermodynamic limit, as [29]

$$b_n = N_L \frac{e^{-n\beta\omega_c}}{n}\prod_{k=1}^{n-1}\frac{k+n\alpha}{k}; \quad b_1 = N_L e^{-\beta\omega_c}. \tag{46}$$

The cluster expansion $\ln Z(\beta,z) = \sum_{n=1}^{\infty} b_n z^n$, as a power series of $ze^{-\beta\omega_c} < 1$, can be summed up

$$\ln Z(\beta,z) = N_L \ln y(ze^{-\beta\omega_c}) \tag{47}$$

where $y(ze^{-\beta\omega_c})$, a function of the variable $ze^{-\beta\omega_c}$, is such that

$$\ln y = ze^{-\beta\omega_c} + \sum_{n=2}^{\infty}\frac{(ze^{-\beta\omega_c})^n}{n}\prod_{k=1}^{n-1}\frac{k+n\alpha}{k}. \tag{48}$$

It obeys [29]

$$y - ze^{-\beta\omega_c}y^{1+\alpha} = 1 \tag{49}$$

and has in turn a power series expansion [38]

$$y = 1 + ze^{-\beta\omega_c} + \sum_{n=2}^{\infty}(ze^{-\beta\omega_c})^n\prod_{k=2}^{n}\frac{k+n\alpha}{k}. \tag{50}$$

From (47) one infers that $Z(\beta,z) = y^{N_L}$ so that [32, 39]

$$Z(\beta,z) = y^{N_L} = 1 + N_L z e^{-\beta\omega_c} + N_L\sum_{N=2}^{\infty}(ze^{-\beta\omega_c})^N\prod_{k=2}^{N}\frac{k+N_L+N\alpha-1}{k}. \tag{51}$$

Clearly, from (51), the N-body partition function Z_N is

$$Z_N = N_L e^{-N\beta\omega_c}\prod_{k=2}^{N}\frac{k+N_L+N\alpha-1}{k}. \tag{52}$$

It is, by construction, positive. Necessarily, α and N_L being given, N has to be such that $N_L + N\alpha \geq 0$. This always is the case as long as N is finite, since N_L scales like the infinite surface of the 2d sample. In the thermodynamic limit, where $N \to \infty$, the condition $N_L + N\alpha \geq 0$ implies for the filling factor

$$\nu \leq -\frac{1}{\alpha}. \tag{53}$$

It is rather striking that the RHS of (53), which has just been derived from the exact computation of the cluster coefficients from the N-body spectrum, is nothing but the critical filling (40) obtained in the mean field approach when the screening is total.

The "degeneracy" associated with N anyons populating the LLL quantum states is, from (52),

$$N_L \prod_{k=2}^{N} \frac{k + N_L + N\alpha - 1}{k} = \frac{N_L}{N!} \frac{(N + N_L + N\alpha - 1)!}{(N_L + N\alpha)!} \tag{54}$$

where a factorial with a negative argument has to be understood as $(-p)! = \lim_{x\to 0}(-p+x)!$.

When $\alpha = 0$, this is the usual Bose counting factor for the number of ways to put N bosons in N_L states

$$\frac{(N + N_L - 1)!}{N!(N_L - 1)!} . \tag{55}$$

When $\alpha = -1$, this is the Fermi counting factor $N_L!/(N!(N_L - N)!)$. If one considers for a moment the statistical parameter to be a negative integer $\alpha \leq -1$, the degeneracy (54) still allows for a combinatorial interpretation [38] : provided again that $N_L + N\alpha \geq 0$, it is the number of ways to put N particles on a circle consisting of N_L quantum states such that there are at least $-\alpha - 1$ empty states in between two occupied states. When $\alpha = -1$, this is nothing but the usual exclusion mechanism for fermions (one fermion at most per quantum state). When $\alpha \leq -1$, i.e., beyond the Fermi point, more and more states are excluded between two filled states. In the case of interest α in $[-1, 0]$, one has a "fractional" exclusion where one can put more than one particle per quantum state according to the fractional α, but not infinitely many as in the Bose case.

The degeneracy (54) originates from the exact N-body spectrum (36). In the case of Haldane statistics as defined in (41), there is no Hamiltonian and no N-body spectrum to begin with. One rather starts from the Bose counting factor (55) and bluntly replaces, in accordance with (41), N_L by $N_L - (N-1)g$ to obtain

$$\frac{(N_L - (N-1)(g-1))!}{N!(N_L - (N-1)g - 1)!} \tag{56}$$

which indeed interpolates, when $g = 1$, to the Fermi counting factor. The degeneracy (56) is similar to (54): if one allows the exclusion parameter g to be an integer, it counts [38] the number of ways to put N particles on a line of finite length consisting of N_L quantum states such that there are at least $g - 1$ empty states in between two occupied states. Up to boundary conditions on the space of available quantum states (periodic versus infinite wall), both counting (54, 56) are identical. In the thermodynamic limit when N becomes large, boundary conditions should not play a role anymore: not surprisingly, starting from (56) and following the usual route of statistical mechanics [40] (saddle-point approximation) leads, in the thermodynamic limit, to the same LLL-anyon thermodynamic potential given by the equations (47) and (49), where the anyonic parameter $-\alpha$ is replaced by the exclusion parameter g.

Note that the grand partition factorization $Z(\beta, z) = y^{N_L}$ in (47) could suggest [41] an interpretation of y as a LLL-anyon grand-partition function for a single quantum state at energy ω_c, on the same footing as, when $\alpha = 0$ or $\alpha = -1$, $y = (1 \mp ze^{-\beta\omega_c})^{\mp 1}$ is indeed the single quantum state grand partition function for a Bose or Fermi gas. This interpretation is not possible for the reason advocated above: it would yield, as soon as α is fractional, negative N-body partition functions. This is clearly impossible: the N-body anyonic system is, except in the Bose and Fermi cases, truly interacting and therefore its statistical mechanics is by no means factorisable to a single-state statistical mechanics.

From (47, 49), the average energy $\bar{E} \equiv -\partial \ln Z(\beta, z)/\partial\beta$ and the average particle number $\bar{N} \equiv z\partial \ln Z(\beta, z)/\partial z$ or, equivalently, the filling factor $\nu = \bar{N}/N_L$, can be computed. ν satisfies

$$y = 1 + \frac{\nu}{1 + \alpha\nu} \tag{57}$$

or, equivalently, using (49),

$$ze^{-\beta\omega_c} = \frac{\nu}{(1 + (1 + \alpha)\nu)^{1+\alpha}(1 + \alpha\nu)^{-\alpha}}\,. \tag{58}$$

When $\alpha \neq 0$ and $\alpha \neq -1$, this equation cannot in general be solved analytically, except in special cases like $\alpha = -1/2$ (semions). The equation of state follows

$$\beta PV = \ln\left(1 + \frac{\nu}{1 + \alpha\nu}\right). \tag{59}$$

In all these equations, it is understood from (53) that $\nu \leq -1/\alpha$. When $\nu = -1/\alpha$, the pressure diverges, a manifestation of the fact that there are as many anyons as possible in the LLL, higher Landau levels being forbidden by construction. One also notes that, for the degenerate LLL gas, the filling factor in (58) is nothing but the mean occupation number n at energy $\epsilon = \omega_c$ and fugacity z. As expected, (58) at $\alpha = 0$ gives the standard Bose mean occupation number $n = ze^{-\beta\epsilon}/(1 - ze^{-\beta\epsilon})$, whereas at $\alpha = -1$ it gives the Fermi mean occupation number $n = ze^{-\beta\epsilon}/(1 + ze^{-\beta\epsilon})$.

The entropy $S \equiv \ln Z(\beta, z) + \beta\bar{E} - (\ln z)\bar{N}$ is (trivially $\bar{E} = \bar{N}\omega_c$ since the N particles are in the LLL)

$$S = N_L\big[(1 + \nu(1 + \alpha))\ln(1 + \nu(1 + \alpha)) - (1 + \nu\alpha)\ln(1 + \nu\alpha) - \nu\ln\nu\big]. \tag{60}$$

It vanishes when $\nu = -1/\alpha$, an indication that the N-body LLL anyon eigenstate is not degenerate at the critical filling. From (36), one infers that the N-body eigenstate of lowest energy has all its one-body orbital momenta quantum numbers $l_i = 0$. It follows from (26) that, in the thermodynamic limit at the critical filling, the LLL-anyon non-degenerate groundstate wavefunction is

$$\psi'(z_1, z_2, ..., z_N; \bar{z}_1, \bar{z}_2, ..., \bar{z}_N) = \prod_{i<j} z_{ij}^{-\alpha} \exp(-\frac{1}{2}\omega_c \sum_{i=1}^{N} z_i\bar{z}_i); \quad \nu = -\frac{1}{\alpha} \tag{61}$$

with total angular momentum

$$L = \frac{N(N-1)}{2\nu}. \tag{62}$$

The pattern in (61) is reminiscent of the Laughlin wavefunctions at FQHE fillings $\nu = 1/(2m+1)$

$$\psi(z_1, z_2, ..., z_N; \bar{z}_1, \bar{z}_2, ..., \bar{z}_N) = \prod_{i<j} z_{ij}^{2m+1} \exp\left(-\frac{1}{2}\omega_c \sum_{i=1}^{N} z_i \bar{z}_i\right); \quad \nu = \frac{1}{2m+1}. \tag{63}$$

On the one hand, Laughlin wavefunctions are fermionic, their filling factors are rational numbers smaller than 1, and they are approximate solutions to the underlying N-body Coulomb dynamics in a strong magnetic field. On the other hand, LLL-anyon wavefunctions are multivalued, their filling factor continuously interpolates between ∞ and 1, and they are exact solutions to the N-body LLL anyon problem. Still, the similarity between (61) and (63) is striking.

Trying to push (61) further beyond the Fermi point eventually up to the Bose point at $\alpha = -2$, one obtains a Bose gas at filling $\nu = 1/2$ with the non-degenerate wavefunction

$$\psi'(z_1, z_2, ..., z_N; \bar{z}_1, \bar{z}_2, ..., \bar{z}_N) = \prod_{i<j} z_{ij}^{2} \exp\left(-\frac{1}{2}\omega_c \sum_{i=1}^{N} z_i \bar{z}_i\right); \quad \nu = \frac{1}{2}. \tag{64}$$

One already knows that the LLL-anyon basis (26) is not interpolating to the complete LLL-Bose basis when $\alpha = -2$. At this point, non LLL N-body eigenstates merge in the LLL ground state to compensate for some missing bosonic quantum numbers -see Figure 3. Clearly, (64) should reproduce, by periodicity, the bosonic non-degenerate wavefunction (61) at $\alpha = 0$, but it does not. On the same footing, when $\alpha = -2$ the critical filling should be bosonic, i.e., $\nu = \infty$, whereas $\nu = 1/2$. The unphysical critical filling discontinuity, ∞ versus 1/2, is yet another manifestation of the missing bosonic quantum numbers. In other words, the very eigenstates which join the LLL ground state at the Bose point $\alpha = -2$ and provide for the missing quantum numbers, have the effect to smooth out the critical filling discontinuity. Still, it has been shown [35] that the stronger the magnetic field B is, the more valid (61) remains closer and closer to $\alpha = -2$. The limit $\alpha \to -2$ is, due to periodicity, the same as the limit $\alpha \to 0$ from above, which can be described as an anti-screening regime. One concludes that close to the Bose point $\alpha = 0$, the critical filling of a LLL-anyon gas is $\nu = \infty$ or $\nu = 1/2$ depending on infinitesimally moving away from the Bose point in the screening regime (the ground state wavefunction is the usual non degenerate bosonic wavefunction), or in the anti-screening regime (the ground state wavefunction is (64)). Again, the Bose point has a somehow singular behavior, a feature already encountered in perturbation theory. Note finally that the occurrence of the $\nu = 1/2$ fraction for the bosonic filling factor in the antiscreening regime is physically challenging: fast rotating Bose-Einstein condensates in the FQHE regime are expected [36] to reach a $1/2$ filling described by the Laughlin-like wavefunction (64).

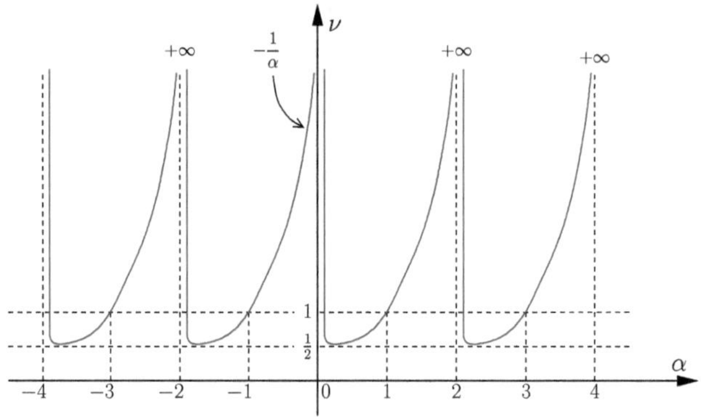

FIGURE 4. The critical LLL-anyon filling curve as a function of α. The critical Bose filling $\nu = \frac{1}{2}$ occurs at the Bose points in the anti-screening regime. The continuity of the critical curve at these points is restored by the non LLL eigenstates joining the LLL ground state.

So far one has been concerned with two-dimensional systems: in the thermodynamic limit, a single particle in the LLL, and, consequently, a gas of LLL-anyons, are two dimensional, as can be seen from the $N_L \simeq V$ scalings[5] of the 1-body LLL partition function $Z_{LLL} = N_L \exp(-\beta\omega_c)$ and the LLL anyon thermodynamic potential (47). Denoting by $\rho_{LLL}(\epsilon) = N_L\delta(\epsilon - \omega_c)$ the 1-body LLL density of states, (47) can be rewritten as

$$\ln Z(\beta, z) = \int_0^\infty \rho_{LLL}(\epsilon) \ln y(ze^{-\beta\epsilon}) d\epsilon. \tag{65}$$

Convincingly, in (65) the one-body dynamics of individual particles is described by the one-body density of states, whereas the LLL statistical collective behavior is encoded in the y function which depends on the statistical parameter α.

One might ask about other integrable N-body systems which would lead to the same kind of statistics. It would be tempting to define a model obeying fractional/exclusion statistics if, its one-body density of states $\rho(\epsilon)$ being given, its thermodynamic potential has the form

$$\ln Z(\beta, z) = \int_0^\infty \rho(\epsilon) \ln y(ze^{-\beta\epsilon}) d\epsilon \tag{66}$$

with

$$y - ze^{-\beta\epsilon} y^{1+\alpha} = 1 \tag{67}$$

[5]In the LLL there is only one quantum number l_i per particle, still the system is 2d.

so that

$$y = 1 + ze^{-\beta\epsilon} + \sum_{n=2}^{\infty} \left(ze^{-\beta\epsilon}\right)^n \prod_{k=2}^{n} \frac{k+n\alpha}{k} . \tag{68}$$

The mean occupation number follows as $n = z\partial \ln y/\partial z$. It obeys

$$y = 1 + \frac{n}{1+\alpha n}; \quad \text{or} \quad n = \frac{y-1}{1-\alpha(y-1)} \geq 0 \tag{69}$$

or, equivalently,

$$ze^{-\beta\epsilon} = \frac{n}{(1+(1+\alpha)n)^{1+\alpha}(1+\alpha n)^{-\alpha}} . \tag{70}$$

One has the duality relation [41]

$$1 = \frac{1}{y} + \frac{1}{\tilde{y}}; \quad \text{where} \quad \tilde{y} - (ze^{-\beta\epsilon})^{-1}\tilde{y}^{1+\frac{1}{\alpha}} = 1 \tag{71}$$

or, equivalently

$$-\alpha n - \frac{1}{\alpha}\tilde{n} = 1 \tag{72}$$

where $\tilde{n}$ is related to $\tilde{y}$ as n to y in (69). The duality relation (71,72) can be interpreted as a particle-hole symmetry relation. Setting $t = ze^{-\beta\omega_c}$, one also has a simple expression [42] for $dn(t)/dt$

$$t\frac{dn}{dt} = n(1+(1+\alpha)n)(1+\alpha n). \tag{73}$$

All these equations have been understood as arising microscopically from the LLL anyon Hamiltonian with one-body density of states $\rho(\epsilon) = \rho_{LLL}(\epsilon)$. It happens that it is possible to find another N-body microscopic Hamiltonian which leads to the thermodynamics (66). Consider, in one dimension, the integrable N-body Calogero model [43] with inverse-square 2-body interactions

$$H_N = -\frac{1}{2}\sum_{i=}^{N} \frac{\partial^2}{\partial x_i^2} + \alpha(1+\alpha)\sum_{i<j} \frac{1}{(x_i - x_j)^2} + \frac{1}{2}\omega^2 \sum_{i=1}^{N} x_i^2 \tag{74}$$

where x_i represents the position of the i-th particle on the infinite 1d line. This model is known to describe particles with nontrivial statistics in one dimension interpolating from Bose ($\alpha = 0$) to Fermi ($\alpha = -1$) statistics. It means that the $1/x^2$ Calogero interaction is purely statistical, without any classical effect on particle motions, up to a overall reshuffling of the particles [44]. The Calogero model remains integrable when, as in (74), a confining 1d harmonic well is added. This is the harmonic Calogero model, whereas the Calogero-Sutherland model [46] would have the particles confined on a circle. The effect of the harmonic well is, as in the LLL anyon case, to lift the thermodynamic limit degeneracy in such a way that the N-body harmonic Calogero spectrum ends up depending on the Calogero coupling constant α

$$E_N = \omega\left[\sum_{i=1}^{N} l_i - \alpha\frac{N(N-1)}{2} + \frac{N}{2}\right]; \quad 0 \leq l_1 \leq l_2 \leq ... \leq l_N. \tag{75}$$

Here the l_i's correspond to the quantum numbers of the 1-d harmonic Hermite polynomials free 1-body eigenstates

$$(\frac{\omega}{\pi})^{1/4}\frac{1}{\sqrt{2^{l_i}l_i!}}e^{-\frac{1}{2}\omega x_i^2}H_{l_i}(\sqrt{\omega}x_i);\quad l_i\geq 0;\quad E=\omega(l_i+\frac{1}{2}). \tag{76}$$

It is remarkable that (75) happens to be again of the form (42) with $\tilde{\omega}=\omega$, $\epsilon_0=\omega/2$. Following the same steps as in the LLL-anyon case, i.e., (44), and using again (45) while taking the thermodynamic limit $\beta\omega\to 0$, the Calogero cluster coefficients rewrite as

$$b_n=\frac{L}{\lambda}\frac{1}{n\sqrt{n}}\prod_{k=1}^{n-1}\frac{k+n\alpha}{k};\quad b_1=\frac{L}{\lambda} \tag{77}$$

where the infinite length of the 1d line has been denoted by L. The cluster expansion can still be resumed using (48) provided the unwanted $1/\sqrt{n}$ term in (77) is properly taken care of. Introducing the 1d plane wave momentum k

$$\frac{1}{\lambda\sqrt{n}}=\frac{1}{2\pi}\int_{-\infty}^{\infty}dke^{-n\beta\frac{k^2}{2}} \tag{78}$$

and denoting the 1-body energy as $\epsilon=k^2/2$, one finally obtains

$$\ln Z(\beta,z)=\int_0^{\infty}\rho_0(\epsilon)\ln y(ze^{-\beta\epsilon})d\epsilon \tag{79}$$

where

$$\rho_0(\epsilon)=\frac{L}{\pi\sqrt{2\epsilon}} \tag{80}$$

is the free 1-body density of states in one dimension. This is not a surprise: in the thermodynamic limit $\omega\to 0$, where $l_i\to\infty$ with $l_i\omega=k_i^2/2$ kept fixed, the Hermite polynomial H_{l_i} becomes a plane wave of momentum k_i.

From (79), one concludes[6] that, in the thermodynamic limit, the Calogero model has indeed a LLL-anyon/exclusion like statistics [45] according to (66) and (67), interpolating, as it should, from a free bosonic 1d gas at $\alpha=0$ to a free fermionic 1d gas at $\alpha=-1$.

It follows that the 2d LLL-anyon and 1d Calogero models, which seem a priori unrelated, do obey the same type of statistics. This is not a coincidence. Looking at their harmonic N-body spectrum (36) and (75), one realizes that, up to an irrelevant zero-point energy, the latter is the $B\to 0$ limit of the former. This remains true in the thermodynamic limit $\omega\to 0$. So, not only (65) and (79) are of the same type, but also, when $B\to 0$, (65) has to become (79). It follows that, necessarily, the 1-body densities of states $\rho_{LLL}(\epsilon)$ and $\rho_0(\epsilon)$ satisfy $\lim_{B\to 0}\rho_{LLL}(\epsilon)=\rho_0(\epsilon)$, i.e.,

$$\lim_{B\to 0}\frac{eBV}{2\pi}\delta(\epsilon-\frac{eB}{2})=\frac{L}{\pi\sqrt{2\epsilon}}, \tag{81}$$

[6] The same conclusion would be reached starting form the Calogero-Sutherland model and taking the corresponding thermodynamic limit, i.e., the radius of the confining circle going to infinity.

a relation which has to be understood as arising in the thermodynamic limit $\omega \to 0$.

To arrive at (81), one could as well consider directly the 1-body harmonic LLL spectrum (31) and harmonic 1d spectrum (76)

$$E = (\omega_t - \omega_c)(l_i + 1) + \omega_c; \quad E = \omega(l_i + \frac{1}{2}). \tag{82}$$

They are such that the latter is the vanishing B limit of the former, so it is the case for the corresponding 1-body partition functions. Taking then[7] the thermodynamic limit $\beta\omega \to 0$, i.e., (45), implies the relation $\lim_{B\to 0} Z_{LLL} = Z_0$, where Z_{LLL} is, as above, the LLL partition function and Z_0 is the free partition function in one dimension. Consequently for the densities of states (the inverse Laplace transforms) the relation (81) follows. This result has its roots in the different energy gaps of the spectra (82) at small ω: in the harmonic LLL case, the gap behaves like $\omega^2/(2\omega_c)$, whereas, in the 1d harmonic case, the gap is ω.

The relation (81) could also have been understood from the 1-body eigenstates themselves. In the limit $B \to 0$, the LLL induced harmonic analytic eigenstates are, from (31),

$$\sqrt{(\frac{\omega^{l_i+1}}{\pi l_i!})} z_i^{l_i} e^{-\frac{1}{2}\omega z_i \bar{z}_i}. \tag{83}$$

There is only one parameter ω left so that the states in (83) can be put in one-to-one correspondence with the Hermite polynomials (76) via the Bargmann transform

$$\sqrt{\omega^{l_i+1}} z_i^{l_i} = \omega \int_{-\infty}^{\infty} dx_i \frac{1}{\sqrt{2^{l_i}}} e^{-\omega(x_i^2 - z_i x_i \sqrt{2} + z_i^2/2)} H_{l_i}(\sqrt{\omega} x_i). \tag{84}$$

From (84) one can infer [47] that the N-body harmonic anyon eigenstates (37) are a coherent state representation of the N-body harmonic Calogero eigenstates.

From all these considerations (thermodynamics, eigenstates,...) it follows that the vanishing magnetic field limit[8] of the LLL-anyon model is the Calogero model itself. It seems paradoxical to consider such a limit in the LLL which assumes a priori a strong magnetic field. Still, doing so, one has dimensionally reduced the 2d anyon model to the 1d Calogero model. This dimensional reduction has a simple geometrical interpretation. The LLL induced harmonic states (83) are localized in the vicinity of circles of radius l_i/ω. In the thermodynamic limit, one has $l_i \to \infty$ with $l_i\omega = k_i^2/2$ kept fixed. It follows that the corresponding 1d Hermite polynomials H_{l_i}, which become in this limit plane waves of momentum k_i, have a radius of localization diverging like k_i^2/ω^2. The dimensional reduction which has taken place consists in going at infinity on the edge of the plane: in the

[7]The order of limits is crucial here: first the limit $B \to 0$, then the thermodynamic limit $\omega \to 0$.
[8]Since one has ended up by taking the limit $B \to 0$, one could have avoided right from the beginning to introduce a B field, and started directly from the harmonic N-body anyon model. What has been done above by taking the limit $B \to 0$ is nothing but to project the harmonic anyon model on the LLL induced harmonic subspace (83) (the B field and its LLL should still be invoked to justify the selection of the LLL quantum numbers in the 2d harmonic basis) and to recognize that the projected harmonic anyon model is the harmonic Calogero model. This relation remains true in the thermodynamic limit $\omega \to 0$.

thermodynamic limit, the Calogero model can be viewed as the edge projection of the anyon model.

The LLL anyon thermodynamics, or, equivalently, the Haldane/exclusion thermodynamics, and the Calogero thermodynamics as well, have been the subject of an intense activity since the mid-nineties. Let us mention their relevance in more abstract contexts, such as conformal field theories [48]. On the experimental side, FQHE edge currents can be modelled by quasiparticles with fractional statistics, which in turn might affect their transport properties such as the current shot noise [49, 42].

4. Minimal Difference Partitions and Trees

Up to now one has been concerned with quantum mechanical models defined by a microscopic quantum Hamiltonian. Both the LLL anyon and Calogero models have been shown to have a thermodynamics controlled by (66) and (67). Let us leave quantum mechanics and address a pure combinatorial problem, the minimal difference partition problem [50]. Consider the number $\rho(E,N)$ of partitions of an integer E into N integer parts where each part differs from the next by at least an integer p and the smallest part is $\geq l$. Usual integer partitions correspond to $p=0$ and $l=1$, whereas restricted partitions, where the parts have to be different, correspond to $p=1$ and $l=1$.

It is known that

$$\sum_E \rho(E,N)x^E = \frac{x^{lN+pN(N-1)/2}}{(1-x)(1-x^2)...(1-x^N)}. \tag{85}$$

The $\rho(E,N)$ generating function $Z(x,z)=\sum_{E,N}^{\infty}\rho(E,N)x^E z^N$ factorizes when $p=0$ or $p=1$

$$p=0,\quad Z(x,z)=\prod_{i=0}^{\infty}\frac{1}{1-x^{l+i}z};\quad p=1,\quad Z(x,z)=\prod_{i=0}^{\infty}(1+x^{l+i}z). \tag{86}$$

In terms of bosons or fermions, (86) is the grand partition function for a bosonic or fermionic gas with fugacity z and, denoting $x=e^{-\beta}$, temperature $T=1/\beta$ where

$$E=\sum_{i=0}^{\infty}n_i(l+i)\quad N=\sum_{i=0}^{\infty}n_i \tag{87}$$

with $n_i=0,1,2,...$ in the Bose case ($p=0$) and $n_i=0,1$ in the Fermi case ($p=1$). Equivalently

$$E=\sum_{i=1}^{N}l_i \tag{88}$$

with $l\leq l_1\leq l_2\leq\ldots\leq l_N$ (Bose) or $l\leq l_1<l_2<\ldots<l_N$ (Fermi).

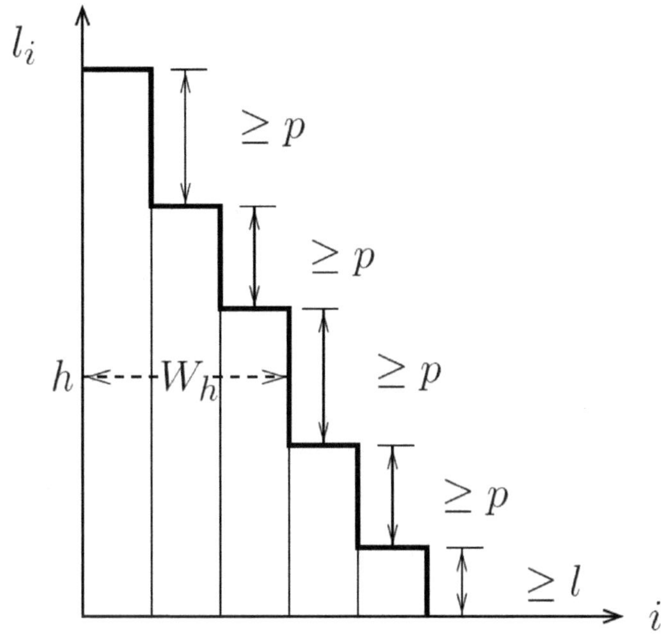

FIGURE 5. A minimal difference partition configuration, or Young diagram. The column heights are such that $(l_i - l_{i+1}) \geq p$ for $i = 1, 2, \ldots, N-1$ and $l_i \geq l$. Their total height is $E = \sum_{i=1}^{N} l_i$. W_h is the width of the Young diagram at height h, i.e., the number of columns whose heights $\geq h$.

When p is an integer ≥ 2, (85) can be regarded as the N-body partition function of an interacting bosonic gas with the N-body spectrum

$$E = \sum_{i=1}^{N} l_i + pN(N-1)/2; \quad l \leq l_1 \leq l_2 \leq \ldots \leq l_N. \tag{89}$$

Clearly, (89) goes beyond the Fermi point $p = 1$ and describes some kind of "superfermions". In contrast to the Bose and Fermi cases, a factorization such as (86) is not possible, due to the interacting nature of (89). One has instead the functional relation

$$Z(x, z) = Z(x, xz) + x^l z Z(x, x^p z) \tag{90}$$

which embodies the combinatorial identity

$$\rho(E, N) = \rho_0\left(E - p\frac{N(N-1)}{2}, N\right) \tag{91}$$

where $\rho_0(E, N)$ stands for the usual partition counting.

One could push [51] this analysis further to p real positive. When $p \in [0,1]$ and $l = 1$, one would obtain a partition problem interpolating between the usual (bosonic) one and the restricted (fermionic) one. It is manifest that, if p is replaced by $-\alpha$, the spectrum (89) coincides, under a rescaling and up to an irrelevant zero-point energy, with the N-body quantum spectrum (75) of the harmonic Calogero model. In a partition problem, one is interested in the large E and N asymptotic behavior of $\rho(E,N)$, which corresponds to the regime $x \to 1$, i.e., $\beta \to 0$. Consider the cluster expansion $\ln Z(x,z) = \sum_{n=1}^{\infty} b_n z^n$. In the limit $\beta \to 0$ one obtains, accordingly to (44) with $\epsilon_0 = l$ and $\tilde{\omega} = 1$,

$$b_n = \frac{1}{\beta}\frac{e^{-nl\beta}}{n^2}\prod_{k=1}^{n-1}\left(1-\frac{pn}{k}\right); \quad b_1 = \frac{1}{\beta}e^{-l\beta}. \tag{92}$$

The limit $\beta \to 0$ should not be confused with the thermodynamic limit in quantum systems. There is no thermodynamic limit prescription like (45). Still, using (48) (with ω_c replaced by l) and taking care of the unwanted $1/n$ factor in (92), one obtains, provided that $ze^{-\beta l} < 1$,

$$\ln Z(\beta, z) = \int_l^{\infty} \ln y(ze^{-\beta\epsilon})d\epsilon \tag{93}$$

with

$$y - ze^{-\beta\epsilon}y^{1-p} = 1. \tag{94}$$

This is again of the form (66) and (67), the statistical parameter $-\alpha$ being replaced by the minimal difference partition parameter p, and the 1-body density of states being the Heaviside function $\rho(\epsilon) = \theta(\epsilon - l)$. The minimal difference partition combinatorics is equivalently described, in the small β limit, by a gas of particles obeying exclusion statistics with a uniform density of states[9].

This correspondence happens to be useful technically: (93) and (94) are the building blocks of the minimal difference partition asymptotics. The average integer $\bar{E} = -\partial \ln Z(\beta,z)/\partial\beta = \int_l^{\infty} n\epsilon d\epsilon$ and the average number of integer parts $\bar{N} = z\partial \ln Z(\beta,z)/\partial z = \int_l^{\infty} n d\epsilon$, are both given in terms of $n = z\partial \ln y/\partial z$, the mean occupation number at "part" ϵ and fugacity z, which satisfies

$$ze^{-\beta\epsilon} = \frac{n}{(1+(1-p)n)^{1-p}(1-pn)^p} \quad \text{with} \quad n \le \frac{1}{p}. \tag{95}$$

One obtains

$$\bar{E} - l\bar{N} = \frac{1}{\beta}\ln Z(\beta,z) \qquad \bar{N} = \frac{1}{\beta}\ln y(ze^{-\beta l}) \tag{96}$$

so that the entropy[10] $S \equiv \ln Z(\beta,z) + \beta\bar{E} - (\ln z)\bar{N}$ rewrites as

$$S = 2\beta\left(\bar{E} - l\bar{N} - \frac{p}{2}\bar{N}^2\right) - \bar{N}\ln(1-e^{-\beta\bar{N}}) \tag{97}$$

[9]There is no microscopic quantum Hamiltonian leading to (93) and (94).

[10]The simple expression in (96) for $\bar{N}$ is possible because of the constant density of states.

with

$$\bar{E} - l\bar{N} - \frac{p}{2}\bar{N}^2 = -\frac{1}{\beta^2}\int_0^{1-e^{-\beta\bar{N}}} \frac{\ln(1-u)}{u} du. \tag{98}$$

Inverting (98) gives β as a function of $\bar{E}$ and $\bar{N}$ so that the entropy S in (97) becomes a function of $\bar{E}$ and $\bar{N}$ only. Doing so, one has a definite information [51] on the asymptotic behavior of $\rho(E,N) \simeq e^{S(E,N)}$ when E and N are large, and also, of $\rho(E) = \sum_{N=1}^{\infty} \rho(E,N)$ when E is large. One obtains a generalization of the Hardy-Ramajunan asymptotics [52] to the minimal difference partition problem. One can also obtain [53] the average limit shape of the Young diagrams associated with the minimal difference partition problem, generalizing the usual partition limit shape [54]. The limit shape at a part of height h depends solely on the statistical function y evaluated at $\epsilon = h$ and at $z = 1$

$$\beta\bar{W}_h = \ln y(e^{-\beta h}) \tag{99}$$

where β scales as $\beta^2 E = \int_0^{\infty} \ln y(e^{-\epsilon}) d\epsilon$.

So far p being a positive integer has insured that the N-body spectrum in (89) is well defined. However, y in (94) is still meaningful when p is a negative integer. It is the $(1-p)$-ary tree generating function, so that the coefficient at order n of its expansion in powers of $ze^{-\beta\epsilon}$ as given in (68) (with $-\alpha$ replaced by p) is the number of ways to build a $(1-p)$-ary tree with n nodes. For example, at $p=-1$, y generates the Catalan numbers associated with binary trees.

Consider, as a toy model [55], the factorized $(1-p)$-ary tree generating function

$$Z(x,z) = \prod_{i=0}^{\infty} y\left(zx^{l+i}\right) \tag{100}$$

where y satisfies (100) with $\epsilon = l+i$. (100) narrows down to (86) when $p=0$ (Bose case). Its combinatorial interpretation is that $\rho(E,N)$ deduced from (100) counts the number of usual partitions of an integer E into N integer parts bigger or equal to l, with an additional degeneracy stemming from the $(1-p)$-tree arborescence when, in a given partition, a part occurs n times. This enlarged degeneracy goes beyond the Bose point to define some kind of "superbosons".

One can analytically continue p to the whole negative real axis. In the large E and N limit, i.e., β smaller and smaller, one encounters a maximal temperature beyond which it is not possible to heat the system. Indeed, from (94) it follows that $y(zx^{l+i})$ in (100) obeys to $y - ze^{-\beta(l+i)}y^{1-p} = 1$, which is well defined only if [39]

$$ze^{-\beta l} < (1-p)^{p-1}(-p)^{-p} < 1. \tag{101}$$

When $z=1$, it defines a dimensionless "Hagedorn temperature"

$$T = \frac{l}{(1-p)\ln(1-p) + p\ln(-p)} \tag{102}$$

just below which E and N become large so that the asymptotic of $\rho(E,N)$ can be addressed.

5. Conclusion

In two dimensions intermediate anyonic statistics interpolating from Bose to Fermi statistics are allowed. Their definition does not involve anything else than the usual concept at the basis of quantum statistics, namely free particles endowed with particular boundary exchange conditions on their N-body wavefunctions. It happens that these boundary conditions have a much richer structure in two dimensions than in three and higher dimensions. This in turn can be understood in terms of the topology of paths in the N-particle configuration space, where non trivial braiding occurs in two dimensions, and not in higher dimensions. A flux-charge composite picture emerges to encode the braiding statistics in physical terms, via topological Aharonov-Bohm interactions and singular magnetic fields.

The anyon model as such is certainly fascinating as far as quantum mechanics is concerned, but it remains an abstract construction whose complexity is daunting. However, when projected onto the LLL of an external magnetic field, the model becomes tractable and, even more, solvable. The LLL set up is clearly adapted to the QHE and to the FQHE physics. Haldane/exclusion statistics, which can be obtained as a LLL-anyon mean-field picture in the screening regime, leads to LLL-anyon thermodynamics.

It would certainly be rewarding if LLL anyons could be relevant experimentally, for example by uncovering some experimental hints at FQHE filling factors of the existence of quasiparticles with anyonic/exclusion statistics. Fractional charges have already been seen in shot noise FQHE experiments [56], but the nontrivial statistical nature of the charge carriers in FQHE edge currents has so far remained elusive in experiments which rely mainly on Aharonov-Bohm interferometry [28]. Note also a recent proposal for the possible experimental tracking of abelian and nonabelian anyonic statistics in Mach-Zehnder interferometers [57].

Finally, on the theoretical side, physical interactions, together with topological anyonic interactions, should also be taken into account in order to produce more realistic models.

Acknowledgements. I would like to thank Alain Comtet and Stefan Mashkevich for past and present collaborations, and for correcting and improving the text. My thanks also to Tobias Paul and Sanjib Sabhapandit for helping me with the figures.

References

[1] J.M. Leinaas, J. Marnheim, *On the theory of identical particles*, Nuovo Cimento **B37** (1977), 1–23. For an earlier work on the subject see: M.G.G. Laidlaw, C.M. de Witt, *Feynman Functional Integrals for Systems of Indistinguishable Particles*, Phys. Rev **D3** (1971), 1375.

[2] For a recent review on spin issues related to statistics see: S. Forte, *Spin in quantum field theory*, Proceedings of the 43th Internationale Univeritastswochen fur Theoretiche Physik, Schladming, Austria (2005). arXiv: hep-th/0507291

[3] G. Moore, N. Seiberg, *Polynomial equations for rational conformal field theories*, Phys. Lett. **B 212** (1988), 451. *Classical and quantum conformal field theory*, Commun. Math. Phys. **123** (1989), 177. G. Moore, N. Read, *Nonabelions in the fractional quantum Hall effect*, Nucl. Phys. **B 360**(2-3) (1991), 362.

[4] A.Yu. Kitaev, *Fault-tolerant quantum computation by anyons*, Ann. Phys. **303** (2003), 2–30.

[5] F. Wilczek, *Magnetic flux, angular momentum, and statistics*, Phys. Rev. Lett. **48** (1982), 1144–1146. *Quantum mechanics of fractional-spin particles*, Phys. Rev. Lett. **49** (1982), 957–959.

[6] Y. Aharonov, D. Bohm, *Significance of electromagnetic potentials in quantum theory*, Phys. Rev. **115** (1959), 485–491. For an earlier work on the subject see: W. Ehrenberg, R.W. Siday, *The Refractive index in electron optics and the principles of dynamics*, Proc. Phys. Soc. London **B62** (1949), 8–21.

[7] R.G. Chambers, *Shift of an electron interference pattern by enclosed magnetic Flux*, Phys. Rev. Lett. **5** (1960), 3.

[8] Tsuyoshi Matsuda, Shuji Hasegawa, Masukazu Igarashi, Toshio Kobayashi, Masayoshi Naito, Hiroshi Kajiyama, Junji Endo, Nobuyuki Osakabe, Akira Tonomura, Ryozo Aoki, *Magnetic field observation of a single flux quantum by electron-holographic interferometry*, Phys. Rev. Lett. **62** (1989), 2519.

[9] W. Siegel, *Unextended superfields in extended supersymmetry*, Nucl. Phys. **B156** (1979), 135–143. J.F. Schonfeld, *A mass term for three-dimensional gauge fields*, Nucl. Phys. **B185** (1981) 157–171. R. Jackiw, S. Templeton, *How super-renormalizable interactions cure their infrared divergences*, Phys. Rev. **D23** (1981), 2291–2304. S. Deser, R. Jackiw, S. Templeton, *Three-dimensional massive gauge theories*, Phys. Rev. Lett. **48** (1982), 975–978. *Topologically massive gauge theories*, Ann. Phys. (N.Y.) **140** (1982), 372–411.

[10] J. McCabe, S. Ouvry, *Perturbative three-body spectrum and the third virial coefficient in the anyon model*, Phys. Lett. **B260** (1991), 113–119.

[11] S. Ouvry, *δ perturbative interactions in the Aharonov-Bohm and anyon models*, Phys. Rev. **D50** (1994), 5296–5299. A. Comtet, S.V. Mashkevich, S. Ouvry, *Magnetic moment and perturbation theory with singular magnetic fields*, Phys. Rev. **D52** (1995), 2594–2597.

[12] C. Manuel, R. Tarrach, *Contact interactions of anyons*, Phys. Lett. **B268** (1991), 222–226.

[13] For a review on the anyon model, see (among others): J. Myrheim, Anyons, Les Houches LXIX Summer School *"Topological aspects of low dimensional systems"* (1998) 265–414.

[14] Y.-S. Wu, *Multiparticle quantum mechanics obeying fractional statistics*, Phys. Rev. Lett. **53** (1984), 111–114.

[15] A.P. Polychronakos, *Exact anyonic states for a general quadratic hamiltonian*, Phys. Lett. **B264** (1991), 362–366. C. Chou, *Multianyon spectra and wave functions*, Phys. Rev. **D44** (1991), 2533–2547. S.V. Mashkevich, *Exact solutions of the*

many-anyon problem, Int. J. Mod. Phys. **A7** (1992), 7931–7942. G. Dunne, A. Lerda, S. Sciuto, C.A. Trugenberger, *Exact multi-anyon wave functions in a magnetic field*, Nucl. Phys. **B370** (1992), 601–635. A. Karlhede, E. Westerberg, *Anyons in a magnetic field*, Int. J. Mod. Phys. **B6** (1992), 1595–1621. S.V. Mashkevich, *Towards the exact spectrum of the three-anyon problem*, Phys. Lett. **B295** (1992), 233–236.

[16] D. Arovas, R. Schrieffer, F. Wilczek, A. Zee, *Statistical mechanics of anyons*, Nucl. Phys. **B251** (1985), 117–126.

[17] A. Comtet, Y. Georgelin, S. Ouvry, *Statistical aspects of the anyon model*, J. Phys. A: Math. Gen. **22** (1989), 3917–3926.

[18] D. Sen, *Spectrum of three anyons in a harmonic potential and the third virial coefficient*, Phys. Rev. Lett. **68** (1992), 2977–2980. M. Sporre, J.J.M. Verbaarschot, I. Zahed, *Anyon spectra and the third virial coefficient*, Nucl. Phys. **B389** (1993), 645–665.

[19] M. Sporre, J.J.M. Verbaarschot, I. Zahed, *Numerical solution of the three-anyon problem*, Phys. Rev. Lett. **67** (1991), 1813–1816. M.V.N. Murthy, J. Law, M. Brack, R.K. Bhaduri, *Quantum spectrum of three anyons in an oscillator potential*, Phys. Rev. Lett. **67** (1991), 1817–1820. M. Sporre, J.J.M. Verbaarschot, I. Zahed, *Four anyons in a harmonic well*, Phys. Rev. **B46** (1992), 5738–5741.

[20] R.K. Bhaduri, R.S. Bhalerao, A. Khare, J. Law, M.V.N. Murthy, *Semiclassical two- and three-anyon partition functions*, Phys. Rev. Lett. **66** (1991), 523–526. F. Illuminati, F. Ravndal, J.Aa. Ruud, *A semi-classical approximation to the three-anyon spectrum*, Phys. Lett. **A161** (1992), 323–325. J.Aa. Ruud, F. Ravndal, *Systematics of the N-anyon spectrum*, Phys. Lett. **B291** (1992), 137–141.

[21] A. Comtet, J. McCabe, S. Ouvry, *Perturbative equation of state for a gas of anyons*, Phys. Lett. **B260** (1991), 372–376.

[22] A. Dasnières de Veigy, S. Ouvry, *Perturbative equation of state for a gas of anyons: Second order*, Phys. Lett. **B291** (1992), 130–136. *Perturbative anyon gas*, Nucl. Phys. **B388** (1992), 715–755.

[23] J. Marnheim, K. Olaussen, *The third virial coefficient of free anyons*, Phys. Lett. **B299** (1993), 267–272. S.V. Mashkevich, J. Marnheim, K. Olaussen, *The third virial coefficient of anyons revisited*, Phys. Lett. **B382** (1996), 124–130. A. Kristoffersen, S.V. Mashkevich, J. Marnheim, K. Olaussen, *The fourth virial coefficient of anyons*, Int. J. Mod. Phys. **A11** (1998), 3723–3747. S.V. Mashkevich, J. Marnheim, K. Olaussen, R. Rietman, *The nature of the three-anyon wave functions*, Phys. Lett. **B348** (1995), 473–480.

[24] R.B. Laughlin, *Quantized Hall conductivity in two dimensions*, Phys. Rev. **B23** (1981), 5632–5633. *Anomalous quantum Hall effect: An incompressible quantum fluid with fractionally charged excitations*, Phys. Rev. Lett. **50** (1983), 1395–1398. *Quantized motion of three two-dimensional electrons in a strong magnetic field*, Phys. Rev. **B27** (1983), 3383–3389. See also: F.D.M. Haldane, *Fractional quantization of the Hall effect: A hierarchy of incompressible quantum fluid States*, Phys. Rev. Lett. **51** (1983), 605–608.

[25] K. von Klitzing, G. Dorda, M. Pepper, *New method for high-accuracy determination of the fine-structure constant based on quantized Hall resistance*, Phys. Rev. Lett. **45** (1980), 494–497. D.C. Tsui, H.L. Störmer, A.C. Gossard, *Zero-resistance state of two-dimensional electrons in a quantizing magnetic field*, Phys. Rev. **B25** (1982),

1405–1407. M.A. Paalanen, D.C. Tsui, A.C. Gossard, *Quantized Hall effect at low temperatures*, Phys. Rev. **B25** (1982), 5566–5569. H.L. Störmer, A. Chang, D.C. Tsui, J.C.M. Hwang, A.C. Gossard, W. Wiegmann, *Fractional quantization of the Hall effect*, Phys. Rev. Lett. **50** (1983), 1953–1956.

[26] D.P. Arovas, R. Schrieffer, F. Wilczek, *Fractional statistics and the quantum Hall effect*, Phys. Rev. Lett. **53** (1994), 722–725. B.I. Halperin, *Statistics of quasiparticles and the hierarchy of fractional quantized Hall states*, Phys. Rev. Lett. **52** (1984), 1583–1586.

[27] H. Kjønsberg, J. Marnheim, *Numerical study of charge and statistics of Laughlin quasiparticles*, Int. J. Mod. Phys. **A14** (1999), 537–557. D. Banerjee, *Topological aspects of phases in fractional quantum Hall effect*, Phys. Lett. **A269** (2000), 138–143.

[28] F.E. Camino, W. Zhou, V.J. Goldman, *Aharonov-Bohm electron interferometer in the integer quantum Hall regime*, arXiv: cond-mat/0503456. *Experimental realization of Laughlin quasiparticle interferometers*, Proc. of EP2DS-17 (Genoa, Italy, 2007). arXiv:0710.1633.

[29] A. Dasnières de Veigy, S. Ouvry, *Equation of state of an anyon gas in a strong magnetic field*, Phys. Rev. Lett. **72** (1994), 600–603.

[30] E. Fermi was the first to introduce an harmonic well confinement to compute thermodynamical quantities: E. Fermi, *Sulla quantizzazione del gas perfetto monoatomico*, Rend. Lincei **3** (1926), 145. In the anyon context, the harmonic well confinement was first used in [17]. See also: K. Olaussen, *On the harmonic oscillator regularization of partition functions*, Trondheim preprint No. 13 (1992).

[31] S.V. Mashkevich, J. Marnheim, K. Olaussen, R. Rietman, *Anyon trajectories and the systematics of the three-anyon spectrum*, Int. J. Mod. Phys. **A11** (1996), 1299–1313.

[32] S. Ouvry, *On the relation between the anyon and the Calogero Models*, Phys. Lett. **B510** (2001), 335.

[33] S. Isakov, G. Lozano, S. Ouvry, *Non abelian Chern-Simons particles in an external magnetic field*, Nucl. Phys. **B552** [FS] (1999), 677.

[34] F.D.M. Haldane, *"Fractional statistics" in arbitrary dimensions: A generalization of the Pauli principle*, Phys. Rev. Lett. **67** (1991), 937–940.

[35] S.V. Mashkevich, S. Ouvry, *The lowest Landau level anyon equation of state in the anti-screening regime*, Phys. Lett. **A310** (2003), 85–94.

[36] N.K. Wilkin, J.M. Gunn, R.A. Smith, *Do attractive bosons condense?*, Phys. Rev. Lett. **80** (1998), 2265.

[37] A. Dasnières de Veigy, S. Ouvry, *One-dimensional statistical mechanics for identical particles : the Calogero and anyon cases*, Mod. Phys. Lett. **B9** (1995), 271.

[38] A.P. Polychronakos, *Probabilities and path-integral realization of exclusion statistics.* arXiv: hep-th/9503077. See also: Generalized statistics in one dimension, Les Houches LXIX Summer School *"Topological aspects of low dimensional systems"* (1998) 415–472.

[39] A.G. Bytsko, *Haldane-Wu statistics and Rogers dilogarithm*, Zap. Nauchn. Semin. POMI **291** (2002), 64–77, J. Math. Sciences **125** (2005), 136–143.

[40] Y.S. Wu, *Statistical distribution for generalized ideal gas of fractional-statistics particles*, Phys. Rev. Lett. **73** (1994), 922–925.

[41] C. Nayak, F. Wilczek, *Exclusion statistics: Low-temperature properties, fluctuations, duality, and applications*, Phys. Rev. Lett. **73** (1994), 2740. S. Chaturvedi, V. Srinivasan, *Microscopic interpretation of Haldane's semion statistics*, Phys. Rev. Lett. **78** (1997), 4316. M.V.N. Murthy, R. Shankar, *Exclusion statistics: A resolution of the problem of negative weights*, Phys. Rev. **B60** (1999), 6517.

[42] G. Gomila, L. Reggiani, *Fractional exclusion statistics and shot noise in ballistic conductors*, Phys. Rev. **B63** (2001), 165404.

[43] F. Calogero, *Solution of a three-body problem in one dimension*, J. Math. Phys. **10** (1969), 2191–2196. *Ground state of a one-dimensional N-body system*, J. Math. Phys. **10** (1969), 2197–2200. *Solution of the one-dimensional N-body problems with quadratic and/or inversely quadratic pair potentials*, J. Math. Phys. **12** (1971), 419–436.

[44] For a recent review on the Calogero model, see (among others): A.P. Polychronakos, Physics and mathematics of Calogero particles. arXiv: hep-th/0607033.

[45] A.P. Polychronakos, *Non-relativistic bosonization and fractional statistics*, Nucl. Phys. **B324** (1989), 597. *Exchange operator formalism for integrable systems of particles*, Phys. Rev. Lett. **69** (1992), 703. S. B. Isakov, *Fractional statistics in one dimension: Modeling by means of $1/x^2$ interaction and statistical mechanics*, Int. J. Mod. Phys. **A9** (1994), 2563. *Generalization of statistics for several species of identical particles*, Mod. Phys. Lett. **B8** (1994), 319. *Bosonic and fermionic single-particle states in the Haldane approach to statistics for identical particles*, Phys. Rev. **B53** (1996), 6585–6590. D. Bernard, Y.-S. Wu, *A Note on statistical interactions and the thermodynamic Bethe ansatz*, (1994). arXiv: cond-mat/9404025. [

[46] B. Sutherland, *Quantum many-body problem in one dimension: Ground state*, J. Math. Phys. **12** (1971), 246–250. *Quantum many-body problem in one dimension: Thermodynamics*, J. Math. Phys. **12** (1971), 251–256.

[47] L. Brink, T.H. Hansson, S. Konstein, M.A. Vasiliev, *The Calogero model anyonic representation, fermionic extension and supersymmetry*, Nucl. Phys. **B401**, issue 3 (1993), 591–612.

[48] K. Schoutens, *Exclusion statistics in conformal field theory spectra*, Phys. Rev. Lett. **79** (1997) 2608–2611. P. Fendley, K. Schoutens, *Cooper pairs and exclusion statistics from coupled free-fermion chains*, J. Stat. Mech. **0207** (2007), 17.

[49] S. Isakov, T. Martin, S. Ouvry, *Conductance and shot noise for particles with exclusion statistics*, Phys. Rev. Lett. **83** (1999), 580.

[50] G.E. Andrews, *The Theory of partitions*, Cambridge University Press, Cambridge (1998). G.E. Andrews, R. Askey, R. Roy, *Special functions*, Encyclopedia of Mathematics and its applications 71, Cambridge University Press, Cambridge (1999).

[51] A. Comtet, S.N. Majumdar, S. Ouvry, *Integer partitions and exclusion statistics*, J. Phys. A: Math. Theor. **40** (2007), 11255.

[52] G.H. Hardy, S. Ramanujan, Proc. London. Math. Soc. **17** (1918), 75.

[53] A. Comtet, S.N. Majumdar, S. Ouvry, S. Sabhapandit, *Integer partitions and exclusion statistics: Limit shapes and the largest part of Young diagrams*, J. Stat. Mech. (2007) P10001.

[54] A.M. Vershik, *Statistical mechanics of combinatorial partitions and their limit shapes*, Functional Analysis and Its Applications **30** (1996), 90.

[55] A. Comtet, S.N. Majumdar, S. Ouvry, S. Sabhapandit, in preparation.

[56] L. Saminadayar, D.C. Glattli, Y. Jin, B. Etienne, *Observation of the e/3 fractionally charged Laughlin quasiparticle*, Phys. Rev. Lett **79** (1997), 162. R. de-Picciotto et al, Nature **389** (1997), 162.

[57] D.E. Feldman, Y. Gefen, A.Yu. Kitaev, K.T. Law, A. Stern, *Shot noise in anyonic Mach-Zehnder interferometer*, Phys. Rev. **B76** (2007), 085333.

Stéphane Ouvry
Université Paris-Sud
Laboratoire de Physique Théorique et Modèles Statistiques
CNRS, UMR 8626
91405 Orsay Cedex, France
e-mail: `stephane.ouvry@lptms.u-psud.fr`

The Spin
B. Duplantier, J.-M. Raimond and V. Rivasseau, Eds., 105–145

Poincaré Seminar 2007

Probing a Single Isolated Electron: New Measurements of the Electron Magnetic Moment and the Fine Structure Constant

Gerald Gabrielse

1. Introduction

For these measurements one electron is suspended for months at a time within a cylindrical Penning trap [1], a device that was invented long ago just for this purpose. The cylindrical Penning trap provides an electrostatic quadrupole potential for trapping and detecting a single electron [2]. At the same time, it provides a right, circular microwave cavity that controls the radiation field and density of states for the electron's cyclotron motion [3].

Quantum jumps between Fock states of the one-electron cyclotron oscillator reveal the quantum limit of a cyclotron [4]. With a surrounding cavity inhibiting synchrotron radiation 140-fold, the jumps show as long as a 13 s Fock state lifetime, and a cyclotron in thermal equilibrium with 1.6 to 4.2 K blackbody photons. These disappear by 80 mK, a temperature 50 times lower than previously achieved with an isolated elementary particle. The cyclotron stays in its ground state until a resonant photon is injected. A quantum cyclotron offers a new route to measuring the electron magnetic moment and the fine structure constant.

The use of electronic feedback is a key element in working with the one-electron quantum cyclotron. A one-electron oscillator is cooled from 5.2 K to 850 mK using electronic feedback [5]. Novel quantum jump thermometry reveals a Boltzmann distribution of oscillator energies and directly measures the corresponding temperature. The ratio of electron temperature and damping rate (also directly measured) is observed to be a fluctuation-dissipation invariant, independent of feedback gain, as predicted for noiseless feedback. The sharply reduced linewidth that results from feedback cooling illustrates the likely importance for improved fundamental measurements and symmetry tests.

Electronic feedback that self-excites the axial oscillation of a single electron in a Penning trap is used to detect spin flips and one-quantum cyclotron excitations [6]. Large, stable, easily detected oscillations arise even in an anharmonic potential. Amplitudes are controlled by adjusting the feedback gain, and frequencies can be made nearly independent of amplitude fluctuations. Quantum jump spectroscopy of a perpendicular cyclotron motion reveals the absolute temperature and amplitude of the self-excited oscillation. The possibility to quickly measure ppb frequency shifts opens the way to improved measurements of the electron magnetic moment.

The new experimental methods make it possible for the first time to use quantum jump spectroscopy of the lowest cyclotron and spin levels for a single-electron quantum cyclotron [7]. The measured electron magnetic moment has an uncertainty that is nearly six times lower than in the past, and the moment is shifted downward by 1.7 standard deviations. The new magnetic moment, with a quantum electrodynamics (QED) calculation, determines the fine structure constant with a 0.7 ppb uncertainty [8, 9] – ten times smaller than for atom-recoil determinations. Remarkably, this 100 mK measurement probes for internal electron structure at 130 GeV.

A big additional reduction in the measurement accuracy is soon to be announced, based upon a new measurement for which the analysis is nearly finished.

2. Quantum Cyclotron

The quantum limit of an electron cyclotron accelerator was demonstrated and reported in collaboration with my student S. Peil [4].

When the cyclotron is cooled to 80 mK, 50 times lower than previously realized with an isolated elementary particle, quantum nondemolition (QND) measurements show that the electron stays in the ground state of its cyclotron motion for hours, leaving only in response to resonant photons deliberately introduced from outside. At higher temperatures, bl ackbody photons are present in sufficient numbers to occasionally excite the electron cyclotron motion. QND measurements show the cyclotron oscillator remains in an excited energy eigenstate for many seconds before making an abrupt quantum jump to an adjacent state. The striking isolation of the electron from its environment is due to a 140-fold cavity-induced suppression of the spontaneous emission of synchrotron radiation. Analysis of the quantum jumps provides a way to measure the temperature of the electron, the average number of blackbody photons, and the spontaneous emission rate. Quantum jump spectroscopy provides a way to precisely measure the frequency separation of the lowest quantum states. A variety of applications are mentioned in conclusion.

The quantum cyclotron provides an unusual opportunity to observe and manipulate long lived states of a harmonic oscillator. When written in terms of raising and lowering operators, the Hamiltonian of the two dimensional cyclotron $H_c = h\nu_c(a^\dagger a + 1/2)$ is formally equivalent to that of the familiar one dimensional harmonic oscillator. The energy eigenstates of the electron cyclotron ($|n = 0\rangle$, $|n = 1\rangle$, ... in Fig. 1a) are often called Landau levels. They are formally equivalent to the familiar number states of the harmonic oscillator, often called Fock states in quantum optics. Though these states are well known to every student of quantum mechanics, the production, observation and use of Fock states in experiments is surprisingly difficult and rare. The unusually high probability $P > 0.999$ to be in the ground state of the quantum cyclotron, and the extremely long lifetime of the Fock states, should make it possible to excite any superposition of the lowest Fock states

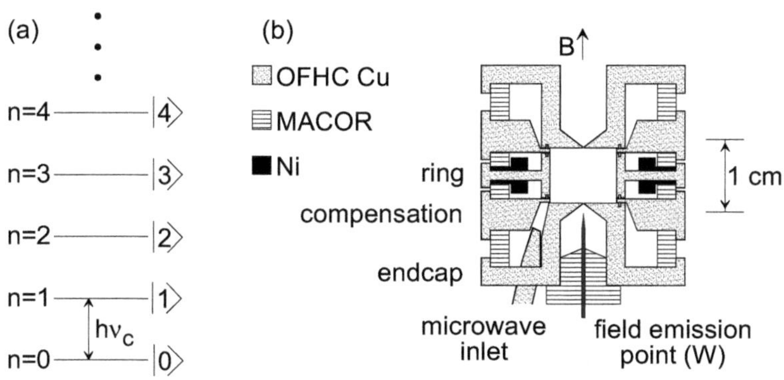

FIGURE 1. (a) Energy levels of the one-electron cyclotron oscillator. (b) Electrodes of the cylindrical Penning trap cavity.

We report the nondestructive observation of Fock states as high as $|n = 4\rangle$. Only zero- and one-photon Fock states, $|n = 0\rangle$ and $|n = 1\rangle$, have previously been observed for a radiation mode of a cavity [10, 11], though efforts are underway to observe two-photon and higher Fock states [12]. A ground state occupation fraction $P = 0.95$ was reported. Vibrational Fock states of a laser-cooled Be^+ ion in a potential well have also been selectively excited, starting from a similar ground state occupation of $P = 0.95$[13]. The formation of these Fock states was deduced destructively, from repeated measurements which transferred the population of identically prepared states to internal energy levels, whose monitored time evolution revealed the original state. Very recently, the $|n = 0\rangle$ and $|n = 1\rangle$ Fock states of neutral atoms oscillating in a one dimensional harmonic well were also observed [14] with $P = 0.92$ for the ground state.

The quantum cyclotron is realized with a single electron stored in a cylindrical Penning trap [4, 2] that is cooled by a dilution refrigerator. The trap cavity (Fig. 1b) is a good approximation to a cylindrical microwave cavity at frequencies up to 160 GHz [15]. Tiny slits (125 μm) in the walls of the cavity make it possible to apply a trapping potential between the central ring electrode and the two flat endcap electrodes. The small slits include quarter wave "choke flanges" to minimize the loss of microwave radiation from the cavity. The potential is made a better approximation to a harmonic potential along the central symmetry axis of the trap by tuning an additional voltage applied to the two compensation electrodes.

Cavity radiation modes that couple to the cyclotron oscillator [15, 16] have quality factors as high as $Q = 5 \times 10^4$. The energy in a 150 GHz mode with this Q value damps exponentially with a 50 ns time constant that is very short compared to all relevant time scales. (The frequency widths of the cavity mode resonances, for example, are much wider than the oscillator's cyclotron resonance width.) The radiation modes of the cavity are thus thermal states with the temperature of the trap cavity. Thermal contact to a dilution refrigerator allows us to adjust the trap temperature between 4.2 K and 70 mK (only to 80 mK when our detector is on.) We detune the frequency of the one-electron cyclotron oscillator away from the radiation modes to decrease the spontaneous emission rate.

Two of the three motions of a trapped electron (charge $-e$ and mass m) in a Penning trap [17] are relevant to this work. Our central focus is upon the circular cyclotron motion, perpendicular to a vertical 5.3 T magnetic field, with cyclotron frequency $\nu_c = eB/(2\pi m) = 147$ GHz and energy levels separated by $h\nu_c$. The Fock states $|n\rangle$, often called Landau states for the particular case of a charged particle in a magnetic field, decay via spontaneous emission to $|n-1\rangle$ at a rate $n\gamma$, where γ is the classical decay rate of the oscillator. In free space for our field, $\gamma = (4\pi\epsilon_o)^{-1} 16\pi^2 \nu_c^2 e^2/(3mc^3) = (94 \text{ ms})^{-1}$. This is the rate that is inhibited by the trap cavity.

The electron is also free to oscillate harmonically along the direction of the vertical magnetic field, $\hat{z}$, at a frequency $\nu_z = 64$ MHz $\approx \nu_c/1000$. We drive this axial motion by applying an oscillatory potential between the ring and an endcap electrode, and detect the oscillatory current induced through a resonant tuned circuit attached between the ring and the other endcap. The electron axial motion damps as energy dissipates in the detection circuit, yielding an observed resonance width of 5 Hz for the driven axial motion. With appropriate amplification and narrow bandwidth detection we are able to measure small (1 Hz) shifts in ν_z. A heterostructure field effect transistor (HFET), constructed with Harvard collaborators just for these experiments, provides the radiofrequency gain that is needed while dissipating only 4.5 μW. The dilution refrigerator had difficulty with the nearly 700 times greater power dissipation (3 mW) of the conventional MESFET used initially.

The cyclotron and axial motions of the electron would be uncoupled except that we incorporate two small nickel rings into the ring electrode of the trap (Fig. 1b). These saturate in and distort the otherwise homogeneous magnetic field. The resulting "magnetic bottle",

$$\Delta\vec{B} = B_2 \left[\left(z^2 - (x^2+y^2)/2 \right) \hat{z} - z(x\hat{x} + y\hat{y}) \right], \tag{1}$$

is similar to but much bigger than what was used to determine an electron spin state [18]. Coupling the combined cyclotron and spin magnetic moment $\vec{\mu}$ to $\Delta\vec{B}$

gives a term in the Hamiltonian that is harmonic in z,

$$V = -\vec{\mu} \cdot \Delta \vec{B} = 2\mu_B B_2 (a^\dagger a + 1/2 + S_z/\hbar) z^2, \tag{2}$$

where μ_B is the Bohr magneton, S_z is the spin operator, and the electron g value is taken to be 2. This V makes ν_z shift in proportion to the energy in the cyclotron and spin motions,

$$\Delta \nu_z = \delta(n + 1/2 + m_s). \tag{3}$$

A one quantum excitation of the cyclotron oscillator shifts the monitored ν_z by $\delta = 2\mu_B B_2/(m\omega_z) = 12.4$ Hz, substantially more than the 5 Hz axial linewidth and the 1 Hz resolution.

The measurement of the cyclotron energy is an example of a QND measurement [19, 20] in that V and H_c commute, $[V, H_c] = 0$. The desirable consequence is that a second measurement of the cyclotron energy at a later time will give the same answer as the first (unless a change is caused by another source). This is not generally true for measurements with a quantum system. For example, measuring the position of a free particle would make its momentum completely uncertain. After additional time evolution a second measurement of the particle's position would give a different outcome.

Five one-hour sequences of QND measurements of the one-electron oscillator's energy are shown in Fig. 2. Each is for a different cavity temperature T, as measured with a ruthenium oxide sensor attached to the ring electrode. Greatly expanded views of several quantum jumps are shown in Fig. 3. Energy quantization is clearly visible, as are the abrupt quantum jumps between Fock states. The upward quantum jumps are absorptions stimulated by the blackbody photons in the trap cavity. The downward transitions are spontaneous or stimulated emissions. Mostly we see the oscillator in its ground state $|n = 0\rangle$, with occasional quantum jumps to excited Fock states. Fig. 3b shows a rare event in which 4.2 K blackbody photons sequentially excite the one-electron cyclotron oscillator to the Fock state $|n = 4\rangle$. It takes of order 2 s of signal averaging for us to ascertain the quantum state of the cyclotron oscillator. This true measurement time is less, being the time required to establish the quantum state in principle. An estimate of this time [21] unfortunately uses assumptions that do not correspond well to the experimental conditions.

We analyze the quantum jumps to measure the temperature of the cyclotron oscillator, T_c. The measured probabilities P_n for occupying Fock states $|n\rangle$, averaged over many hours, are shown to the right in Fig. 2 for each cavity temperature. The measured P_n fit well to the Boltzmann factors $P_n = Ae^{-nh\nu_c/kT_c}$ which pertain for thermal equilibrium, demonstrating that averaged over hours the oscillator is in a thermal state. The fit determines T_c. Measurements with this "quantum Boltzmann thermometer" (solid points in Fig. 4a) shows that T_c is equal to the cavity temperature T; the cyclotron oscillator is in thermal equilibrium with the blackbody photons in the cavity. The solid points in Fig. 4b show the measured average quantum number superimposed upon the curve $\bar{n} = [e^{h\nu_c/kT} - 1]^{-1}$ which pertains for an oscillator in thermal equilibrium at the measured cavity temperature T. For temperatures of 4.2 K, 1 K and 80 mK, $\bar{n}$ varies dramatically from 0.23, to 9×10^{-4}, to 6×10^{-39}.

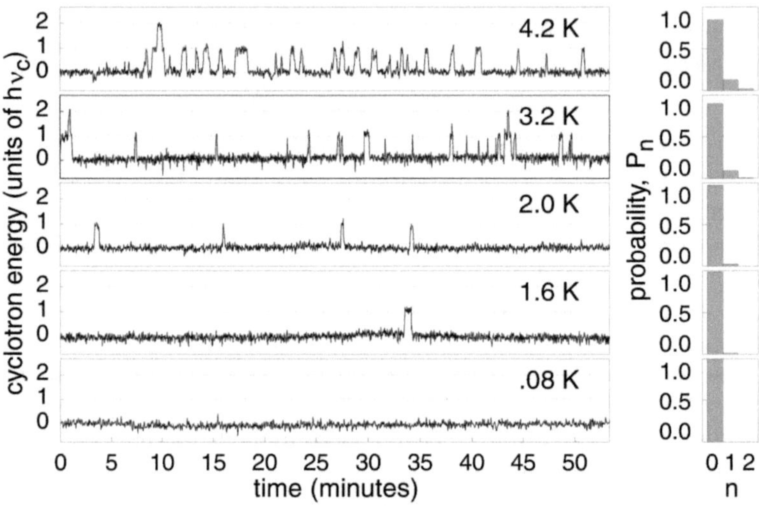

FIGURE 2. Quantum jumps between the lowest states of the one-electron cyclotron oscillator decrease in frequency as the cavity temperature is lowered.

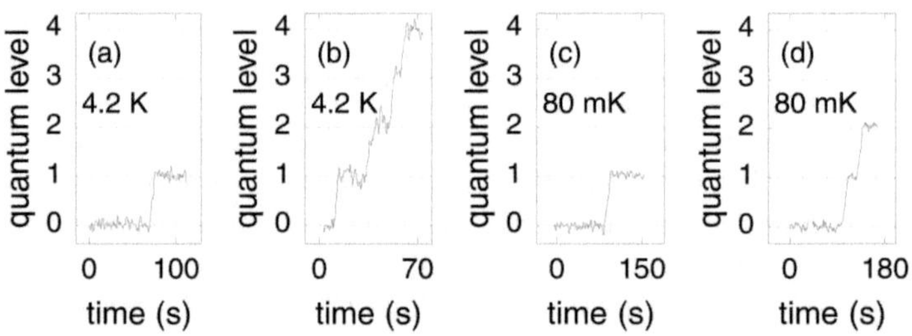

FIGURE 3. Excitations to excited Fock states which are stimulated by 4.2 K blackbody photons in (a) and (b), and by an externally applied microwave field in (c) and (d).

Below 1 K the oscillator resides in its ground state for so long (we estimate 10^{32} years for 80 mK) that it is difficult to directly measure the oscillator temperature T_c. The best we can do is to establish that at some confidence level C, this temperature is below a limit given by $kT_c \leq h\nu_c/\ln[1 - \gamma t/\ln(1 - C)]$ if we observe no excitation for time t. When no excitation is observed for $t = 5$ hours, for example, we establish that $T_c < 1.0$ K at the $C = 68\%$ confidence level. For temperatures below 1 K, blackbody photons have been essentially eliminated, and the one-electron cyclotron oscillator is virtually isolated from its environment.

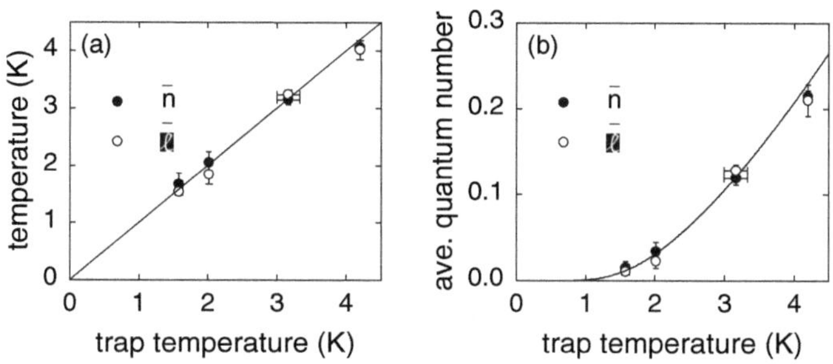

FIGURE 4. (a) The oscillator temperatures deduced from the measured occupation times in each number state (solid points) and deduced from the transition rates (open points) are compared to the temperature of a ruthenium oxide thermometer attached to a trap electrode. (b) Measured average values $\bar{n}$ and $\bar{\ell}$ as a function of cavity temperature.

We can separately measure the rate Γ_{abs} for the upward jumps (corresponding to stimulated absorption), and the rate Γ_{em} for downward jumps (corresponding to stimulated and spontaneous emission together). For $T = 1.6$ K, Fig. 5 shows a histogram of the dwell times in $|n = 0\rangle$ in (a) and for $|n = 1\rangle$ in (b). Both histograms decrease exponentially, indicating random processes, so the fitted lifetimes $(\Gamma_{abs})^{-1}$ and $(\Gamma_{em})^{-1}$ are just the average values of the dwell times. The rates for stimulated emission from $|n\rangle$ to $|n-1\rangle$ and for stimulated absorption from $|n-1\rangle$ to $|n\rangle$ are expected to be equal by the principle of detailed balance. Thus the spontaneous emission rate is simply the difference between the observed emission rate and the observed absorption rate, $\gamma = \Gamma_{em} - \Gamma_{abs}$. At $T = 1.6$ K (Fig. 5) the measured stimulated absorption rate is negligibly smaller so that $\gamma^{-1} \approx \Gamma_{em}^{-1} = 13$ s.

Comparing the 13 s spontaneous emission lifetime that is measured with the 94 ms expected for free space shows that spontaneous emission of synchrotron radiation is strongly suppressed. The 140-fold inhibition is due to the copper trap cavity that encloses the electron oscillator [22]. By adjusting the magnetic field, the frequency of the cyclotron oscillator is tuned away from resonance with the radiation modes of the trap cavity. The electron oscillator then couples only very weakly to the modes of the radiation field, and spontaneous emission is suppressed. We would not otherwise be able to signal average sufficiently to observe the quantum jumps so distinctly, nor would the excited Fock states persist so long.

The measured emission and absorption rates determine the average number $\bar{\ell}$ of resonant blackbody photons within the cavity. Quantum electrodynamics indicates that stimulated emission from $|n\rangle$, and stimulated absorption into $|n\rangle$, both

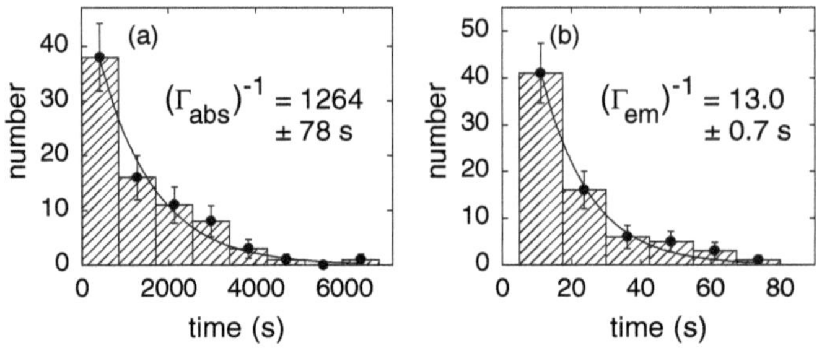

FIGURE 5. Histograms of the dwell times preceding stimulated absorption from $n = 0$ to $n = 1$ in (a), and for spontaneous and stimulated emissions from $n = 1$ to $n = 0$ in (b), both for $T = 1.6$ K. Dwell times less than 5 s are excluded since short dwell times are obscured by detection time constants.

have the same rate given by $\bar{\ell}n\gamma$. Applied to $n = 1$, this means that $\Gamma_{abs} = \bar{\ell}\gamma$ and $\Gamma_{em} = (1 + \bar{\ell})\gamma$. The average number of blackbody photons in terms of measurable quantities is thus given by $\bar{\ell} = \Gamma_{abs}/(\Gamma_{em} - \Gamma_{abs})$. The measured open points in Fig. 4b agree well with the expected curve $\bar{\ell} = [e^{h\nu_c/kT} - 1]^{-1}$, and $\bar{n} = \bar{\ell}$ as predicted. Fitting to the measured $\bar{\ell}$ gives an independent measurement of the temperature of the cavity (open points in Fig. 4a). These agree well with the directly measured cavity temperature.

Extremely precise quantum jump spectroscopy of the lowest levels of the quantum cyclotron should become possible with blackbody photons eliminated from the trap cavity. Quantum jumps (e.g. Fig. 3c-d) will take place only when externally generated microwave photons are introduced into the trap cavity, increasing in rate as the drive frequency is swept through resonance. One challenge is that the z^2 term in the magnetic bottle (Eq. 1) not only couples ν_z to the cyclotron energy (Eq. 3) as is desired for good detection sensitivity. It also shifts the cyclotron frequency in proportion to the axial energy E_z with $\Delta\nu_c = \delta\, E_z/(h\nu_z)$. The measured distribution of cyclotron frequencies shows that the current axial detector heats the axial motion of the electron to 17 K, well above the 80 mK temperature of the trap and cyclotron motion. However, the long lifetime of the first excited Fock state should make it possible to introduce microwave photons while the axial motion is cooled to 80 mK, before turning on the axial detector to observe whether a cyclotron excitation has been made.

In conclusion, a quantum cyclotron is demonstrated using one electron in a cylindrical Penning trap cavity. QND measurements of quantum jumps between cyclotron Fock states shows that the temperature of the cyclotron motion tracks the cavity temperature where this can be measured, from 4.2 K to 1.6 K. At 80 mK the electron is 50 times colder than previously demonstrated for an isolated elementary particle. Blackbody photons are completely absent and the cyclotron remains in its quantum ground state. The jumps also show that the Fock states are long lived; the cavity suppresses the spontaneous emission of synchrotron radiation 140-fold.

The quantum cyclotron is so well prepared in its ground state, and so well isolated from its environment, that it may be possible to excite any desired superposition of excited states, to probe the nature of decoherence and quantum measurement. Quantum jump spectroscopy offers the prospect to measure the frequency between the lowest Fock states (and spin states) with the exquisite precision required to significantly improve the very accurate measurement of the the electron magnetic moment and the fine structure constants, as illustrated in following sections. A better lepton CPT test, comparing the magnetic moments of the electron and positron, should be possible, along with a better measurement of the proton-to-electron mass ratio.

The work of this section was done as part of the Ph.D. work of S. Peil, with early experimental contributions from K. Abdullah and D. Enzer. Support came from the NSF with some assistance from the ONR.

3. Feedback Cooling

Feedback cooling of the axial motion of a single suspended electron in a cylindrical Penning trap was initially reported in collaboration with my students B. D'Urso, B. Odom and D. Hanneke [5].

At a time when the importance of feedback for reducing amplifier noise was already recognized [23], Kittel described the theory and limits of "noiseless" feedback damping [24]. Feedback damping has been applied in subsequent decades to a variety of oscillatory systems including an electrometer[25], a torsion balance [26], a mechanical gravity gradiometer [27], a laboratory rotor [28], a vibration mode of an optical mirror [29], and to the stochastic cooling of particle beams [30]. The possible application of Kittel's "noiseless" feedback to trapped particles was mentioned [31], as was the relevance of the limitations he discussed [32] to proposed stochastic cooling of trapped antiprotons [33]. Using feedback to improve measurements is an active area of current research [34].

This section describes the feedback cooling of the simplest of oscillators – one with demonstrated potential for fundamental measurements. A one-electron oscillator is cooled from 5.2 K to 0.85 K. A unique feature is that this classical oscillator's temperature and damping rate are both determined absolutely by measuring frequencies. A novel feature is that quantum jump thermometry (utilizing quantum electron cyclotron motion orthogonal to the cooled classical oscillation) directly displays the Boltzmann distribution of oscillator energies [4]. The measurements reveal cooling to an ideal, noiseless feedback limit that is characterized by a fluctuation-dissipation invariant. Noise added by the active feedback electronics limits the lowest temperature attained.

The observed narrowing of an electron's cyclotron resonance line, with similar narrowing of the "anomaly" resonance [17] at the difference of its spin and cyclotron frequencies, will allow higher precision measurements of these frequencies and more precise systematic studies. The higher accuracy determination of these frequencies expected as a result could enable better measurements of the magnetic moments of the electron and positron, an improved determination of the fine structure constant, an improved CPT test with leptons, and a better measurement of the proton-to-electron mass ratio.

The oscillation cooled with feedback is that of a single electron along the central symmetry axis ($\hat{z}$) of a cylindrical Penning trap [1, 2] (Fig. 6). The trap electrodes are biased so the electron oscillates in a harmonic potential well ($\sim z^2$) with frequency $\nu_z = 64.787$ MHz. The z^4 well distortion is tuned out by adjusting the potential on small, orthogonalized compensation electrodes [1].

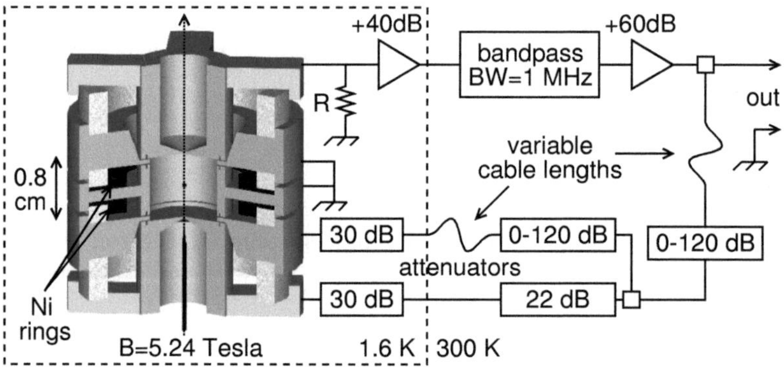

FIGURE 6. Representation of trap and high frequency electronics used for feedback cooling. Static potentials applied to suspend the electron at the trap center are not shown.

We treat the one-electron oscillator as a charge attached to a massless spring, focussing upon potentials and currents that oscillate near ν_z, while ignoring the additional static trapping potentials always applied to the trap. Oscillatory potentials applied to either of the two endplate electrodes (Fig. 7) drive the electron oscillator. The electron motion, in turn, induces a current I to flow through R, a resistance due to unavoidable loss in an attached amplifier and inductor. The inductor (in parallel to R but not shown) tunes out trap capacitance (e.g. between the plates).

With no feedback (Fig. 7a), the induced current I removes energy from the electron oscillator at the familiar rate I^2R, with the result that the damping rate $\Gamma \propto R$. The proportionality constant depends upon the electron charge, the electron mass and the geometry of the trap [17]. Measurements to be discussed show that the electron oscillator is weakly damped (i.e. $\Gamma/2\pi \ll \nu_z$) with $\Gamma/2\pi = 8.4$ Hz.

The random thermal fluctuations of electrons within R, in thermal equilibrium at temperature T, produce a fluctuating Johnson-Nyquist noise potential

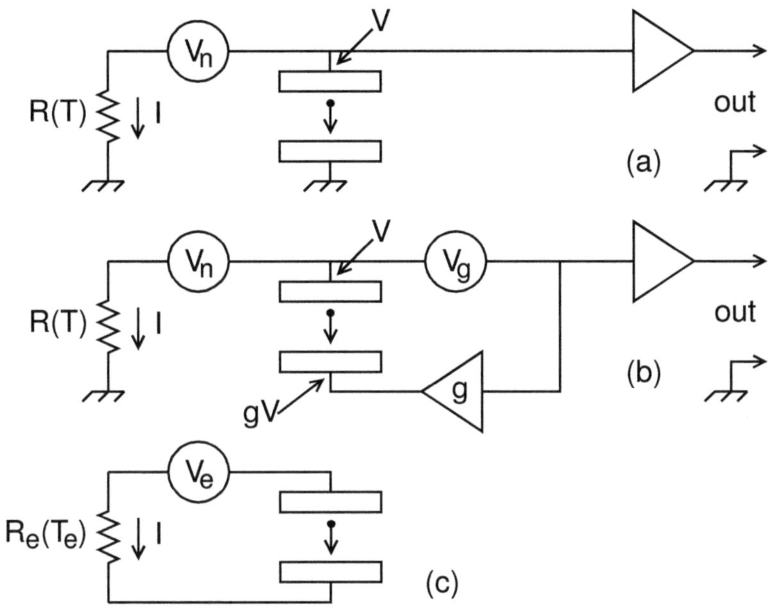

FIGURE 7. Conceptual circuit without (a) and with (b) feedback. For ideal noiseless feedback, $V_g = 0$. For the electron, circuits (b) and (c) are equivalent.

[35, 36] V_n. This frequency independent white noise, with

$$\overline{V_n^2} = 4kTR\Delta\nu \tag{4}$$

in a frequency bandwidth $\Delta\nu$, drives the electron. This noise and the induced current both contribute to the voltage on the upper plate, $V = V_n + IR$. A sensitive HEMT (high electron mobility transistor) amplifier [37] amplifies V so it can be detected.

The measured power spectrum for V (Fig. 8a) has a constant baseline due to the Johnson noise. The current induced by the noise-driven electron produces a notch in this flat spectrum at ν_z; the angular frequency width of this notch is the damping rate Γ. The notch is most easily understood if the oscillating charge is represented as a familiar electrical oscillator, an inductor ℓ and a capacitor c in series, connected between the plates. On resonance at ν_z the electron acts as an electrical short between the plates since the reactances of the ℓ and c cancel. The notch has the characteristic Lorentzian shape of a damped harmonic oscillator. The observed noise cancellation is not perfect (i.e. the dip does not go perfectly to zero power in Fig. 8a) because of amplifier noise, trap potentials that are not perfectly stable, and residual trap anharmonicity.

When the amplifier is on, as it must be for feedback to be applied, measurements to be discussed show that R is at a temperature of 5.2 K. This is higher

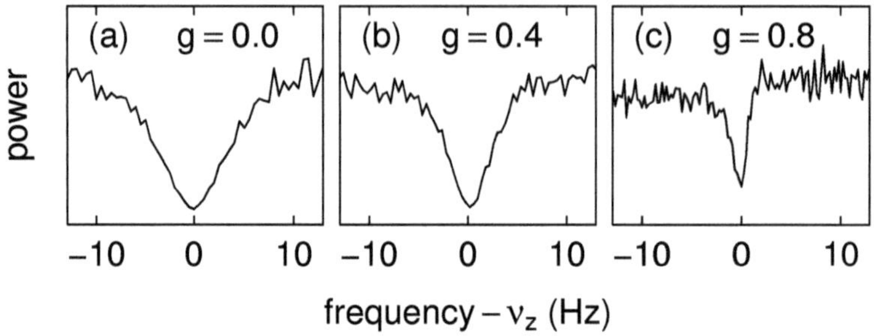

FIGURE 8. Oscillator damping rate Γ_e (the width of the notch in the white Johnson noise) without feedback (a) and when decreased using feedback (b)-(c).

than the 1.6 K of the trap apparatus (maintained by thermal contact to a pumped ^{4}He system), despite the minimal 420 μW power dissipation of the current-starved HEMT, and heroic efforts to thermally anchor the HEMT at 1.6 K.

Feedback is applied as shown conceptually in Fig. 7b. The fluctuating upper plate voltage V is fed back to the lower plate with feedback gain, g. A more complete representation (Fig. 6) shows amplifiers, attenuators and variable cable lengths used to adjust the feedback phases. Correctly phased feedback to two electrodes, rather than just to the bottom plate in the conceptual Fig. 7b, applies feedback to the electron while cancelling feedback to the amplifier. Feedback to the amplifier would modify its properties [23], perhaps improving particle detection in some situations [38], but would complicate the relationship between feedback gain, electron temperature and electron damping.

For the electron, the effect of feedback is equivalent to the circuit in Fig. 7c, with R_e and T_e chosen to make the motion-induced potential and the fluctuation potential across the plates the same as for Fig. 7b.

To determine R_e (and hence the damping rate $\Gamma_e \propto R_e$) we insist that electron motion induces the same potential difference across the plates. Equating $IR - gIR$ for Fig. 7b with IR_e for Fig. 7c yields $R_e = (1-g)R$, and an electron damping rate

$$\Gamma_e = (1-g)\Gamma. \tag{5}$$

When $g = 0$ we recover the damping rate Γ for no feedback. When $g = 1$ the electron oscillator is undamped.

To determine the effective temperature T_e we insist that the electron see the same noise fluctuations across the plates in Figs. 7b and 7c. Equating $V_n - gV_n$ for Fig. 7b with V_e for Fig. 7c, yields

$$T_e = (1-g)T. \tag{6}$$

We recover the resistor temperature T when there is no feedback ($g = 0$). The temperature decreases as the feedback gain is increased. We shall see that noise added in the feedback process prevents attaining 0 K as $g \to 1$.

The fluctuations (characterized by a temperature) and the dissipation (characterized by a damping rate) are related for ideal, noiseless feedback by a fluctuation-dissipation invariant [24],

$$T_e/\Gamma_e = T/\Gamma. \tag{7}$$

Noiseless feedback with gain $g < 0$ increases the damping rate but at the expense of also increasing the temperature and fluctuations. Noiseless feedback cooling, with $0 < g < 1$, decreases the temperature, but at the expense of reducing the damping rate. The advantage of a reduced T_e is to reduce deleterious effects of axial fluctuations upon other electron motions, as we will illustrate with a reduced cyclotron linewidth.

Real feedback amplifiers add fluctuations V_g that increase T_e above the the ideal Eqs. 6-7, and reduce the depth of the observed Lorentzian noise notch. Equating the fluctuations across the plates for Figs. 7b-c yields

$$T_e = T\left[1 - g + \frac{g^2}{1-g}\frac{T_g}{T}\right]. \tag{8}$$

T_g is a feedback "noise temperature" such that $\overline{V_g^2}/\overline{V_n^2} = T_g/T$. The relative depth of the Lorentzian notch in the observed noise power,

$$F = 1 - (1-g)^{-2}(1 + T/T_g)^{-1}, \tag{9}$$

is the ratio of this noise power on and off resonance.

T_e initially drops linearly with g increasing from zero as in the ideal case (Eq. 6). (An example is the function fit to measured temperatures in Fig. 10a, discussed later.) T_e then rises rapidly as $g \to 1$, the limit of an undamped oscillator driven by feedback noise.

The lowest temperature is $T_e(min) \approx 2\sqrt{T_g T}$, for $T_g \ll T$, is at an optimal feedback gain $g \approx 1 - \sqrt{(T_g/T)}$, and our amplifier has $T_g \approx 40$ mK. Meanwhile, the deep notch ($F \approx 1$ for $g = 0$) goes to essentially no notch at all ($F \approx 0$) at the gain that minimizes T_e. Damping remains but we cannot measure its rate by this method.

The temperature T_e of the effective damping resistance is important because the electron axial oscillation comes into thermal equilibrium at T_e. Averaged over many axial damping times Γ^{-1}, the probability that the oscillator has energy between E_z and $E_z + dE_z$ goes as the Boltzmann factor, e^{-E_z/kT_e}.

Remarkably, we can directly measure this Boltzmann distribution, and hence T_e, using quantum jump thermometry. The quantum jumps [4] are between the ground and first excited states of the electron's cyclotron motion in a 5.24 Tesla magnetic field directed along the electron's axial oscillation (Fig. 6). Compared to the rapid 146.7 GHz cyclotron motion the axial motion is adiabatic. It is unaffected by a single quantum cyclotron excitation except for the tiny shift of ν_z (Eq. 11) that we discuss next. The cyclotron damping lifetime is extended to 15 seconds (from 0.1 seconds for free space) using a trap cavity that inhibits spontaneous emission [4].

The coupling of cyclotron and axial motion comes from the small "magnetic bottle" gradient [17] from two small nickel rings (Fig. 6). The electron sees a magnetic field that increases slightly as z^2 as it moves away from the center of the trap in its axial oscillation. This coupling shifts the cyclotron frequency by a measured [4] $\delta = 12$ Hz for every quantum of axial excitation,

$$\Delta\nu_c = \delta(E_z/h\nu_z). \tag{10}$$

The axial frequency shifts by the same amount,

$$\Delta\nu_z = \delta(E_c/h\nu_c), \tag{11}$$

for every quantum of cyclotron excitation. Both tiny shifts are used for the quantum jump thermometry.

A Boltzmann distribution of axial energies, owing to Eq. 10, makes an associated distribution of cyclotron frequencies, given that the axial damping time is longer than the time associated with the noise fluctuations of the axial frequency [17]. A cyclotron driving force at frequency ν excites a quantum jump between the ground and first excited cyclotron states with a probability

$$P(\nu) \sim \begin{cases} 0, & \nu < \nu_c \\ e^{-\frac{\nu_z}{\delta}\frac{h(\nu-\nu_c)}{kT_e}}, & \nu > \nu_c, \end{cases} \tag{12}$$

provided that the jumps happen more rapidly than the one per hour stimulated by black body photons in the 1.6 K trap cavity.

To determine whether a quantum jump has taken place we look for the corresponding axial frequency shift (Eq. 11). We do not simply observe the center frequency of a noise dip (Fig. 8), though this would likely suffice. Instead, before a cyclotron excitation decays (in typically 15 s), we observe the electron's response to a strong axial drive for the 1 second needed to measure $\Delta\nu_z$ and determine the cyclotron state.

The measurement cycle starts with 0.5 s of magnetron sideband cooling [17] to keep the electron near the center axis of the trap. Feedback cooling is then applied for 6 seconds, with the cyclotron drive at ν turned on for the last 2 of these seconds. The axial drive to determine $\Delta\nu_z$ and the cyclotron state is applied next, along with more magnetron cooling. The cyclotron state is read out once each second until the ground state is observed for 2 s. The cycle then repeats.

The measured cyclotron lineshapes (Fig. 9) narrow significantly as the feedback gain increases. Each shows the characteristic Boltzmann distribution that signifies thermal equilibrium. Each is fit to Eq. 12 to determine the equilibrium axial temperature, T_e. The lowest observed $T_e = 850$ mK (Fig. 9c) is a substantial reduction of the 5.2 K realized without feedback.

The measured axial temperature (Fig. 10a) decreases linearly as g increases from 0, as predicted in Eqs. 6 and 8. There is a good fit to Eq. 8, including the

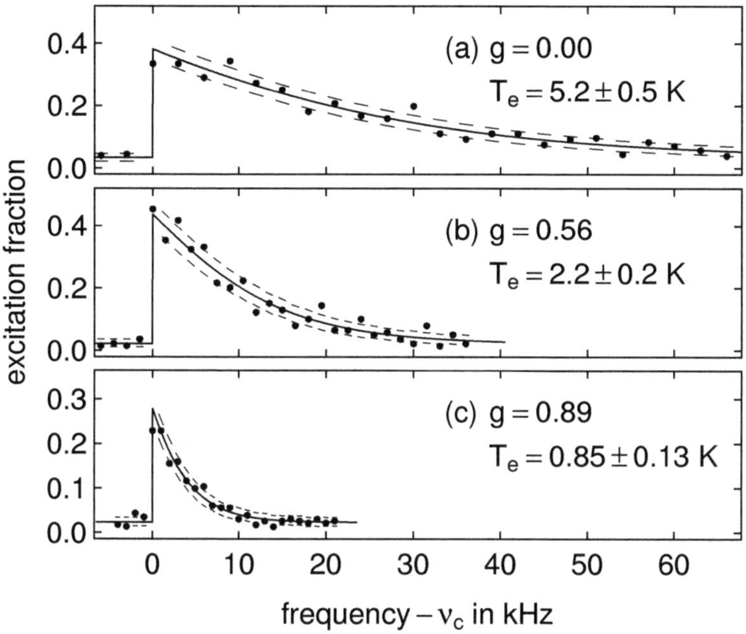

FIGURE 9. Cyclotron resonances show a Boltzmann distribution of axial energies that decreases as feedback gain g in increased. Dashed lines bound the 68% confidence area.

rapid increase for $g \to 1$ which corresponds to a nearly undamped system being driven by the noise added in the feedback signal. It is difficult to fix g accurately enough to measure points on this rapid rise.

The damping rate Γ_e, the width of a noise dip (e.g. Fig. 8), is measured directly (Fig. 10b). The damping rate decreases linearly with increasing g as predicted in Eq. 5. The vanishing dip width and the instabilities mentioned earlier keep us from measuring Γ near to $g = 1$.

Because we directly measure T_e (characterizing fluctuations) and Γ_e (characterizing dissipation) we can compare their ratio (Fig. 10c) to the fluctuation-dissipation invariant that pertains for noiseless feedback (Eq. 7). The measured ratio is invariant and is at the ideal limit, though we expect that it would rise above the ideal limit if we could measure it for feedback gains closer to unity.

In conclusion, feedback cooling to the noiseless limit is demonstrated with the simplest of oscillators. Characterization of the cooling of a one-electron oscillator is direct and complete because both fluctuations and dissipation are directly and absolutely determined by frequency measurements. In addition, sharply narrowed cyclotron lineshapes present the possibility of much more accurate measurements

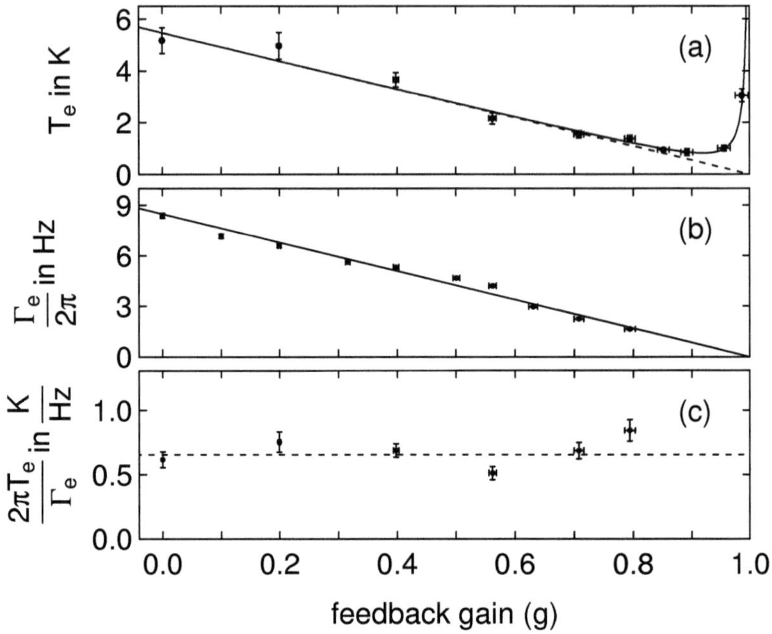

FIGURE 10. Axial temperature (a), damping rate (b) and the fluctuation-dissipation invariant (c) as a function of feedback gain. The dotted line in (c) is the weighted average.

of the electron cyclotron frequency, with similar line narrowing and accuracy improvement expected for the electron "anomaly" resonance [17]. Better measurements of these frequencies for a single trapped electron and positron opens the way to better measurements of their magnetic moments, a more accurate value of the fine structure constant, a more precise test of CPT invariance for leptons, and an improved proton-to-electron mass ratio.

The work of this section was supported by the NSF, the ONR, and the AFOSR. B.D. was also supported by the Fannie and John Hertz Foundation.

4. First One-Particle Self-Excited Oscillator

The initial demonstration of the use of electronic feedback to produce a one-particle self-excited oscillator was done in collaboration with my students B. D'Urso, R. Van Handel, B. Odom and D. Hanneke [6]

The harmonic motion of an oscillator can be excited and sustained with a driving force derived from its own oscillation. A wide range of macroscopic oscillators are operated as self-excited oscillators (SEO), from the electromechanical clock [39] and its ubiquitous quartz successors, to the nanomechanical cantilevers

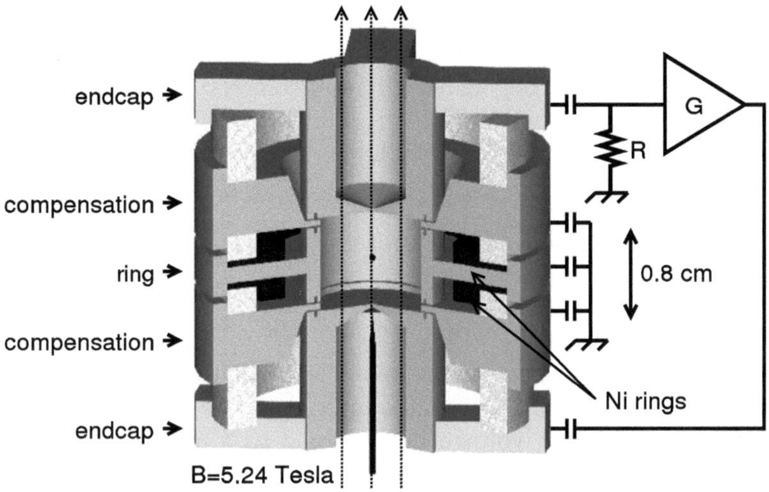

FIGURE 11. The vertical oscillation of a trapped electron, shown within a cutaway of a cylindrical Penning trap, induces a voltage across resistor R that is amplified and fed back to drive the oscillation. Unavoidable trap capacitance in parallel to R is tuned out at ω_z with a parallel inductor.

used in atomic force microscopes [40] and sensitive electrometers [41]. A microscopic SEO is more difficult to realize because such small signals and driving forces are involved. The possibility of realizing a one-ion SEO in a Paul trap was once discussed [31], and self-driven feedback cooling of a one-electron oscillator has been realized [5].

In this section we demonstrate a microscopic, one-particle SEO for the first time. The axial motion of a single electron suspended in a Penning trap is driven by an electric field derived from the current that its motion induces in an electrical circuit. The principal challenge is in stabilizing the electron's oscillation amplitude, an amplitude measured here using quantum jump spectroscopy of a perpendicular cyclotron motion. The frequency stability and the signal-to-noise allow detection of a 5 parts in 10^{10} frequency shift in a few seconds – a sensitivity that allows the detection of a one quantum change in the electron cyclotron energy and an electron spin flip. Likely applications are improved measurements of the electron, positron, proton and antiproton magnetic moments.

The oscillation which is self-excited is that of a single electron (charge $-e$ and mass m) along the central axis ($\hat{\mathbf{z}}$) of a cylindrical Penning trap [1] (Fig. 11) maintained at either 0.1 or 1.6 K. A ring electrode at potential $-V_0$ with respect to grounded endcaps generates a potential on the z-axis,

$$\Phi(z) = \frac{V_0}{2}\left[C_2\frac{z^2}{d^2} + C_4\frac{z^4}{d^4} + C_6\frac{z^6}{d^6} + \cdots\right], \tag{13}$$

where $d = 0.35$ cm indicates the trap size. The C_k are determined by trap geometry and by the potential V_c applied to small compensation electrodes (Fig. 11) to adjust C_4 and C_6. An "orthogonalized" trap geometry [1] makes C_2 essentially independent of V_c.

A drive force $F_d(t)$ and a damping force $-m\gamma_z\dot{z}$ yield

$$\ddot{z} + \gamma_z\dot{z} + [\omega_z(A)]^2 z = F_d(t)/m. \tag{14}$$

The angular oscillation frequency $\omega_z(A)$ is

$$\frac{\omega_z(A)}{\omega_z} \approx 1 + \frac{3C_4}{4C_2}\left(\frac{A}{d}\right)^2 + \frac{15C_6}{16C_2}\left(\frac{A}{d}\right)^4, \tag{15}$$

for small $(C_4/C_2)^2$. It depends weakly upon the oscillation amplitude A [17], and $\omega_z = \sqrt{eV_0C_2/(md^2)}$ pertains for small amplitudes.

The motion induces a voltage proportional to $\dot{z}$ across a $Q \approx 600$ tuned circuit (R in Fig. 11). Energy dissipated in R damps the motion. The signal is amplified with a high electron mobility transistor (Fujitsu FHX13LG) anchored to the cryogenic environment, and operated at a very low current to minimize trap heating. For the two realizations of the SEO that we will describe, typically 420 μW is used to detect the comparator SEO operated at 1.6 K, and only 12 μW for the DSP SEO at 0.1 K. Some amplified signal is phase shifted and fed back to the opposite endcap to drive the SEO. The rest is fourier transformed to determine its amplitude and the SEO oscillation frequency.

The feedback produces a force $F_d(t) = Gm\gamma_z\dot{z}$. Feedback cooling of the electron motion takes place if $G < 1$ [5]. Self-excitation occurs in principle when the feedback cancels the damping, for unit feedback gain $G = 1$. Any noise will cause amplitude diffusion and energy growth, however. Also, if G differs even slightly from unity, A will either decrease or increase exponentially.

A stable and useful SEO thus requires a fixed oscillation amplitude A_o, arranged using an amplitude-dependent gain $G(A)$ that decreases with increasing A near $G(A_o) = 1$. This gain in Eq. (14) yields

$$\dot{A} = -\tfrac{1}{2}\gamma_z A\,[1 - G(A)] \tag{16}$$

for the time evolution of the amplitude [42]. In practice, the gain-control system may average the signal for a time τ before determining A. Eq. (16) is valid if $\tau \gg 1/\omega_z$ and $1/\tau$ is much larger than the resulting self-excited oscillator linewidth.

We demonstrate two methods of stabilizing the amplitude of a SEO – passing the feedback drive through a comparator (Fig. 12a) and employing a fast digital signal processor or DSP (Fig. 12b). The first was realized at $\omega_z/(2\pi) = 64$ MHz and the second at $\omega_z/(2\pi) = 200$ MHz for reasons not related to this demonstration. The comparator is simpler, but the DSP is the more flexible option that can be made much more immune to noise. For both demonstrations the technical noise added by the feedback amplifier is so small [5] that we neglect it in our analysis. We were unable with the electrodes of our trap to realize a third method – applying the signal induced on one electrode to a second electrode (not the image of the

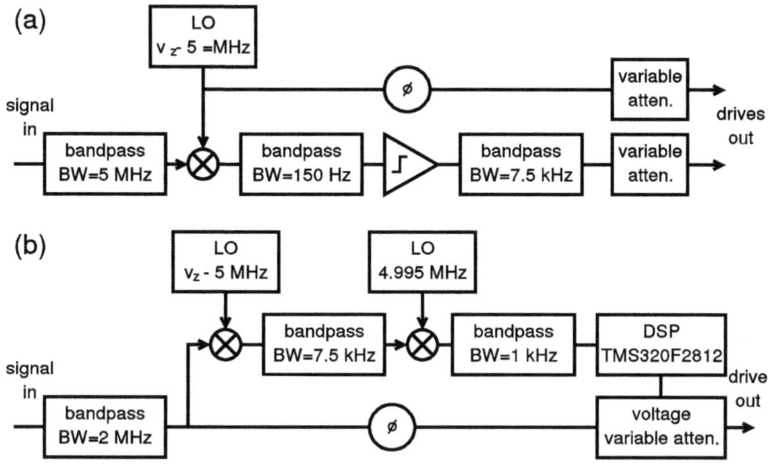

FIGURE 12. Overview of the comparator (a) and DSP (b) feedback used to obtain amplitude stabilization. Phase shifters are labeled with ϕ.

first under $z \to -z$) to make the effective feedback gain decrease with oscillation amplitude.

A noiseless feedback drive passing through a comparator generates a fixed oscillation amplitude A_o. Thus $G(A) = A_o/A$ and Eq. (16) together yield

$$\dot{A} = -\tfrac{1}{2}\gamma_z(A - A_o). \tag{17}$$

The amplitude A damps exponentially to A_o; the time constant $\gamma_z/2$ is the same as for damping without feedback. Noise injected into a comparator softens its gain response, limiting the gain at low A [43]. Narrow band filters (Fig. 12a) to reduce the noise are thus essential. A big challenge is in adjusting the trapping potential to keep the shifting oscillation frequency centered on the filters. (The drives to two electrodes in Fig. 12a are effectively one drive insofar as the electron mixes the two frequencies [17].)

We program the DSP chip to calculate a running Fourier transform of the amplified induced signal, and to adjust the feedback gain as a cubic function of the largest transform amplitude, so

$$\dot{A} = -\tfrac{1}{2}\gamma_z[a_1(A - A_o) + a_2(A - A_o)^2 + a_3(A - A_o)^3]. \tag{18}$$

For this demonstration only the linear term is used, with $a_2 = a_3 = 0$. The effective bandwidth (related to the fourier transform bin width) is 8 Hz, but the "filter" is always centered on the oscillation frequency. No square wave is generated so no filtering of harmonics is required.

Fig. 13 shows that $\omega_z(A_o) - \omega_z$ depends approximately quadratically upon A_o^2 as predicted by Eq. 15 for $A \ll d$. A_o is varied by changing the gain for three different V_c. A_o is determined from the size of the induced signal to which it is

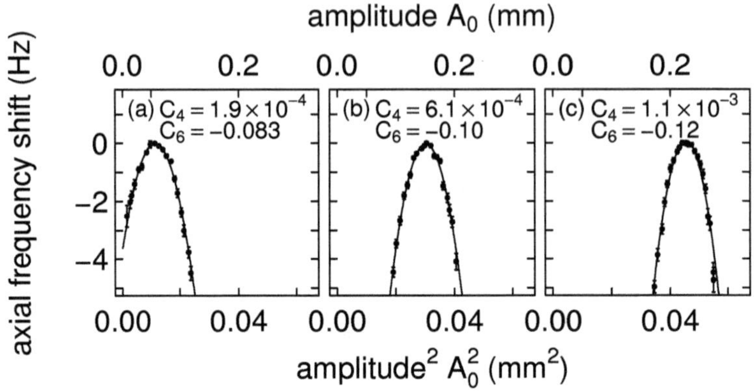

FIGURE 13. The axial frequency measured as a function of the square of the axial amplitude A_o using the comparator SEO.

proportional, with a proportionality constant that will be discussed shortly. Fits of Eq. 15 to the measurements in Fig. 13 allow us to determine and adjust C_4 and C_6 with unprecedented accuracy.

One consequence is that extremely small frequency shifts can be quickly detected with the SEO. In a given averaging time, a frequency can typically be measured to the familiar limit provided by the uncertainty principle, divided by the signal-to-noise ratio (S/N) [44]. A large induced S is possible due to the large oscillation amplitudes, illustrated in Fig. 13. The effect of amplitude fluctuation noise N (driven by thermal fluctuations in the detection resistor) is particularly small if the oscillator amplitude is stabilized at a maxima (e.g. Fig. 13) caused when the effects of C_4 and C_6 of opposite sign cancel. An oscillator is locally harmonic at such maxima, with the oscillation frequency insensitive to small, noise-driven, amplitude fluctuations, despite the large oscillation in an anharmonic potential. Fig. 14a shows the standard deviation of repeated frequency measurements as a function of averaging time. With only four seconds of averaging time a 5 parts in 10^{10} (0.5 ppb) shift in ω_z can be measured – a substantial improvement on any other method.

In principle, a fixed frequency drive could sustain a large oscillation in an anharmonic potential. In practice, however, if the oscillator frequency changes suddenly the oscillation could be lost. Also, a fixed frequency drive cannot generally build up a large oscillation amplitude in the first place since the oscillator shifts out of resonance with the drive as the oscillation amplitude increases.

The considerable advantage of a SEO is that its self-derived drive always stays resonant, even if its oscillation frequency changes suddenly. It also stays resonant while the oscillation amplitude builds up to a large value, during which time the oscillation frequency is shifting. Typically our SEO is excited in less than 1 second.

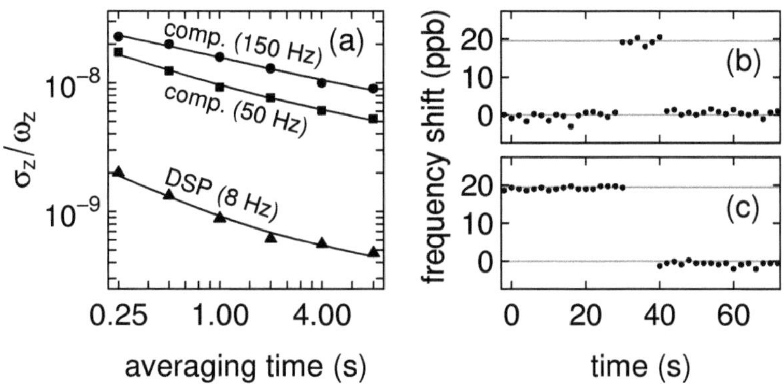

FIGURE 14. Fractional standard deviation of repeated frequency measurements for SEO with indicated bandwidth (a). Small shifts in the frequency of a 200 MHz SEO indicate a one-quantum cyclotron excitation (b) and a separate spin flip (c). The SEO is off while a drive is applied to flip the spin, giving the gap in (c).

A calibration of the axial oscillation amplitude – using quantum jump spectroscopy of an orthogonal cyclotron motion – is next. Noise applied to the oscillator gives a distribution of axial energies E_z about the stable oscillation energy E_o, amplitude A_o and phase ϕ_o that pertain for no noise. A reservoir at temperature T_z, weakly coupled to the axial motion, gives a distribution [45]

$$P(E_z; E_o, T_z) = \frac{1}{kT_z} e^{-\frac{E_z+E_o}{kT_z}} I_0\left(\frac{2\sqrt{E_z E_o}}{kT_z}\right), \tag{19}$$

where I_0 is a modified Bessel function. No feedback drive gives $E_o \to 0$ and a Boltzmann distribution of E_z.

A simple derivation verifies this distribution and highlights the assumptions. For small fluctuations from A_o the oscillation can be taken as harmonic at angular frequency $\omega_o = \omega_z(A_o)$, with $E_o = \frac{1}{2}m\omega_o^2|A_o e^{i\phi_o}|^2$. Noise alone would drive the oscillator into thermal equilibrium, to a Boltzmann distribution of energies $E_n = \frac{1}{2}m\omega_o^2|A_n e^{i\phi_n}|^2$, where ϕ_n is a random oscillation phase. The oscillation amplitude A_z and phase ϕ_z due to independent feedback and noise drives is the superposition $A_z e^{i\phi_z} = A_o e^{i\phi_o} + A_n e^{i\phi_n}$ for a harmonic oscillation. The combined effect of feedback and noise for a particular ϕ_n arises from the distribution of the total amplitude

$$\tilde{P}(A_z e^{i\phi_z}) \propto e^{-\frac{E_n}{kT_z}} = e^{-\frac{m\omega_o^2|A_z e^{i\phi_z} - A_o e^{i\phi_o}|^2}{2kT_z}}. \tag{20}$$

The probability distribution of $E_z = \frac{1}{2}m\omega_o^2|A_z e^{i\phi_z}|^2$ in Eq. 19 is the average of this distribution over random ϕ_z.

Remarkably, quantum jump spectroscopy directly measures Eq. 19 and thus determines A_o and T_z. The quantum jumps [4] are between the ground and first excited states of cyclotron motion at frequency $\nu_c = 148$ GHz in a $B = 5.24$ Tesla magnetic field $B\hat{\mathbf{z}}$. A small "magnetic bottle" gradient $\mathbf{\Delta B} \sim z^2\hat{\mathbf{z}}$ [46] from two nickel rings (Fig. 11) weakly couples the cyclotron (or spin) magnetic moment μ to the axial motion, adding a coupling term that goes as $\mu\Delta B \sim \mu z^2$ to the Hamiltonian.

The corresponding small addition to the oscillator's restoring force, $\sim \mu z$, shifts the observed axial oscillation frequency in proportion to μ. Our frequency resolution makes it possible to observe that ω_z shifts by δ for a single quantum excitation from the cyclotron ground state (Fig. 14b). The probability P_c that a cyclotron driving force at a frequency ν causes a quantum jump [4] thus becomes measurable. For the 200 MHz oscillator, $\delta = 3.9$ Hz. For the 64 MHz oscillator, $\delta = 12$ Hz.

The second consequence of the magnetic bottle coupling is that the magnetic field averaged over an axial oscillation changes with oscillation energy, shifting ν_c by δ for every quantum of axial energy. The quantum jump spectrum,

$$P_c(\nu; E_o, T_z) \propto P(\hbar\omega_z(\nu - \nu_c)/\delta; E_o, T_z), \tag{21}$$

thus reveals the distribution of axial energies of Eq. 19.

Figs. 15a-c show examples for the three axial oscillation amplitudes that result for the trap settings of Fig. 13a-c. The extracted temperature T_z (Fig. 15d) is independent of oscillation amplitude. The detection amplifier makes T_z hotter than the ambient temperature, emphasizing that either this amplifier must be off, or feedback cooling must be applied, [5] to achieve low axial temperatures. Fig. 15e shows that the extracted $E_o \sim A_o^2$ is proportional to the induced signal power, which can be measured in only seconds. The quantum jump spectroscopy in Fig. 15 which calibrates this signal took about 40 hours.

The one-electron SEO allows such good detection of small frequency shifts that a likely application is the measurement of electron and positron magnetic moments – to provide the most accurate direct lepton CPT test, and the most accurate determination of the fine structure constant α. Figs. 14b-c illustrate the detection of a one-quantum cyclotron excitation and a spin flip. Quantum jump spectroscopy – measuring the number of quantum jumps as appropriate drive frequencies are changed – could provide the first fully quantum measurement of these moments.

Averaging the frequency of a one-antiproton ($\overline{\mathrm{p}}$) SEO over a long time, to detect extremely small δ, may make possible the long-time goal [47] of measuring the $\overline{\mathrm{p}}$ magnetic moment. Improving the 0.3% current accuracy [48] by a factor of a million or more seems conceivable. The needed ν_c can already be measured to such an accuracy [49]. Measuring the needed spin precession frequency requires observing a $\overline{\mathrm{p}}$ spin flip. The δ that would signal such a flip is proportional to $\mu/\sqrt{m}$ for a particle with magnetic moment μ and mass m. The challenge is that μ for the $\overline{\mathrm{p}}$ is 658 times smaller than that of the electron, and $\sqrt{m}$ is 43 times larger, so that δ is a daunting 3×10^4 smaller than that of an an electron in the same trap.

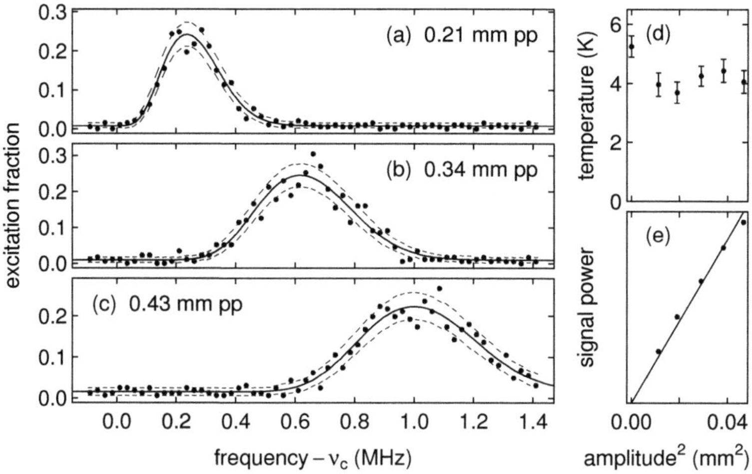

FIGURE 15. (a-c) Measured quantum jump spectra (points), fits to $P_c(\nu; E_o, T_z)$ (solid) and 68% confidence limits (dashes) for the comparator SEO conditions of Fig. 13a-c. (d) T_z from the fits. (e) Induced signal (in arbitrary units) is proportional to A_o^2 from fits.

Fortunately, the size of the frequency shift δ can be increased, since δ is proportional to an apparatus factor $\beta M/(d\sqrt{V_o})$ [17]. For example, making the ring electrode in the trap of Fig. 11 out of iron rather than copper would increase the product of a relative geometry factor β and the magnetization M for the magnetic material (and hence δ) by a factor of 16. Substantial additional increases could come from reducing the trap size and potential, d and V_o, limited by the extent to which this makes a more anharmonic axial oscillation. The fractional stability required in the trapping potential goes as $\mu\beta M/V_o$ and seems possible. To avoid broadened resonances, spin flips and cyclotron excitations would be made in a trap without a magnetic gradient, then transferred to a detection trap with a large magnetic gradient, as in measurements of magnetic moments of bound electrons [50].

In conclusion, self-excitation is demonstrated with the simplest of microscopic oscillators - a single electron suspended in a Penning trap. Both a comparator and a DSP are used to stabilize large, easily observed oscillations that are much larger than noise-driven fluctuations. Despite the anharmonic trap potential, with the right choice of feedback gain, the SEO rapidly excites itself to a large oscillation that is locally harmonic – with an oscillation frequency largely independent of amplitude fluctuations. It maintains the large oscillation even when its oscillation frequency shifts suddenly. The great signal-to-noise ratio observed with the SEO makes it possible to detect small frequency shifts quickly. The SEO could thus enable better measurements of the electron and positron magnetic moments. It may also make it possible to detect antiproton spin flips for the first time, thereby opening the way to greatly improved measurements of the antiproton magnetic moment.

The work of this section was supported by the NSF. B.D. was also supported by the Fannie and John Hertz Foundation, and D.H. by the ARO.

5. New Measurement of the Electron Magnetic Moment

The discussion of a new measurement of the electron magnetic moment in this section is based a recent report [7] that was published in collaboration with my students, B. Odom, B. D'Urso and D. Hanneke.

Measurements of the electron magnetic moment ($\boldsymbol{\mu}$) probe the electron's interaction with the fluctuating vacuum of QED, and also probe for possible electron substructure. As an eigenstate of spin $\mathbf{S}$, the electron (charge $-e$ and mass m) has $\boldsymbol{\mu} \propto \mathbf{S}$,

$$\boldsymbol{\mu} = -g \frac{e\hbar}{2m} \frac{\mathbf{S}}{\hbar}. \tag{22}$$

The g-value is a dimensionless measure of the moment[1], with the dimensions and approximate size given by the Bohr magneton, $e\hbar/(2m)$. If the electron was a mechanical system with an orbital angular momentum, then g would depend upon the relative distributions of the rotating charge and mass, with $g = 1$ for identical distributions. (Cyclotron motion of a charge in a magnetic field B, at frequency $\nu_c = eB/(2\pi m)$, is one example.) A Dirac point particle has $g = 2$. QED predicts that vacuum fluctuations and polarization slightly increase this value. Electron substructure [51] would make g deviate from the Dirac/QED prediction (as quark-gluon substructure does for a proton).

Measurements of the electron g have a long history [52, 53], with a celebrated measurement [18] providing the accepted value [54] since 1987. The new g has a six times smaller standard deviation and is shifted by 1.7 standard deviations (Fig. 16a). A one-electron quantum cyclotron [4], cavity-inhibited spontaneous emission [22], a self-excited oscillator (SEO) [6], and a cylindrical Penning trap [1] contribute to the extremely small uncertainty. For the first time, spectroscopy is done with the lowest cyclotron and spin levels of a single electron fully resolved via quantum non-demolition (QND) measurements [4], and a cavity shift of g is directly observed.

What can be learned from the more accurate electron g? The first result beyond g itself is the fine structure constant, $\alpha = e^2/(4\pi\epsilon_0\hbar c)$, determined from g and QED with ten times smaller uncertainty compared to any other method [8]. This fundamental measure of the strength of the electromagnetic interaction is a crucial ingredient in our system of fundamental constants [54]. Second, the most demanding test of QED continues to be a comparison of measured and calculated g-values, and the way is now prepared for a ten times more stringent test. Third, even

---

[1]The "g" used here and in what follows should not be confused the feedback gain in Section 3 which unfortunately is referred to with the same symbol.

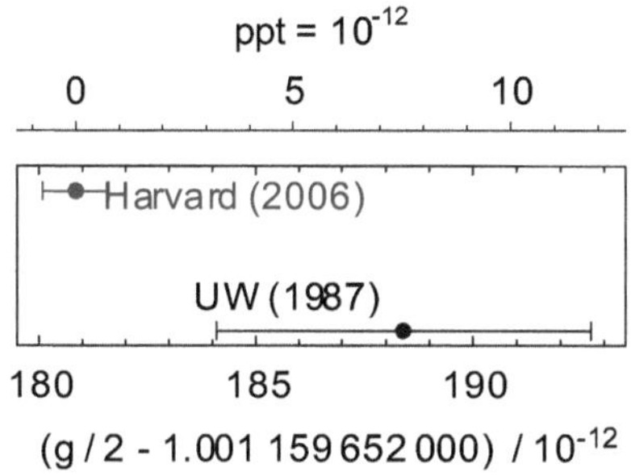

FIGURE 16. Measurements of the electron g-value.

though muon g-values [55] have nearly 1000 times larger uncertainties compared to the electron g, heavy particles (possibly unknown in the standard model) make a contribution that is relatively much larger for the muon. However, the contribution is small compared to the calculated QED contribution which depends on α and must be subtracted out. The electron g provides α and a confidence-building test of the required QED.

The g-value determines the spin frequency $\nu_s = \frac{g}{2}\nu_c$ for a free electron in a magnetic field $B\hat{\mathbf{z}}$. To weakly confine the electron, an electric quadrupole potential, $V \sim 2z^2 - \rho^2$, is added, with $\boldsymbol{\rho} = x\hat{\mathbf{x}} + y\hat{\mathbf{y}}$. Optimal biasing of the electrodes (Fig. 17a) of an orthogonalized cylindrical Penning trap [1] minimizes an undesired z^4 term. The electron-trap system has four eigenfrequencies. The spin and trap-modified cyclotron frequencies are approximately equal at $\nu_s \approx \bar{\nu}_c \approx 149$ GHz. A harmonic axial oscillation along $\mathbf{B}$ is at $\bar{\nu}_z \approx 200$ MHz, and an orthogonal circular magnetron oscillation is at $\bar{\nu}_m \approx 134$ kHz. The latter three frequencies are shifted by the unavoidable leading imperfections of a real Penning trap – harmonic distortions of the quadrupole potential, and a misalignment of the electrode axis and $\mathbf{B}$ [56]. Silver trap electrodes were used after the nuclear paramagnetism of copper electrodes caused unacceptable temperature-dependent fluctuations in $\mathbf{B}$ near 100 mK.

The spin motion is undamped, being essentially uncoupled from its environment [17]. The cyclotron motion would damp in ~ 0.1 s via synchrotron radiation in free space. This spontaneous emission is greatly inhibited in the trap cavity (to 6.7 s or 1.4 s here) when $\mathbf{B}$ is tuned so $\bar{\nu}_c$ is far from resonance with cavity radiation modes [22, 17]. Blackbody photons that would excite the cyclotron ground state are eliminated by cooling the trap and vacuum enclosure below 100 mK with a dilution refrigerator [4]. (Thermal radiation through the microwave inlet makes < 1 excitation/hr.) The axial motion, damped by a resonant circuit, cools below 0.3 K (from 5 K) when the axial detection amplifier is off for crucial periods. The magnetron motion radius is minimized with axial sideband cooling [17].

For the first time, g is deduced from observed transitions between only the lowest of the spin ($m_s = \pm 1/2$) and cyclotron ($n = 0, 1, 2, \ldots$) energy levels

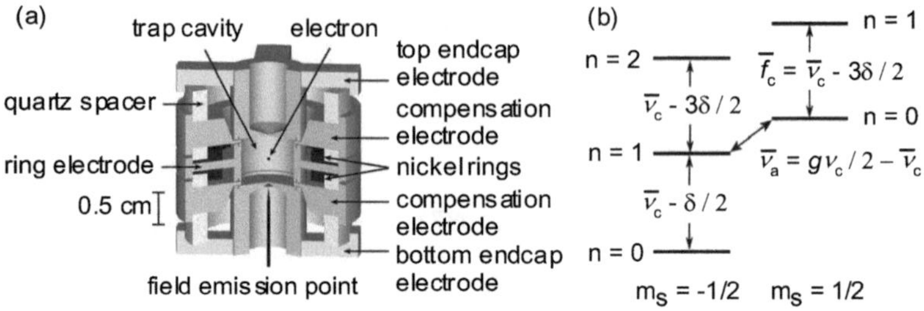

FIGURE 17. Cylindrical Penning trap cavity used to confine a single electron and inhibit spontaneous emission (a), and the cyclotron and spin levels of an electron confined within it (b).

(Fig. 17b),

$$E(n, m_s) = \frac{g}{2} h\nu_c m_s + (n + \tfrac{1}{2})h\bar{\nu}_c - \tfrac{1}{2}h\delta(n + \tfrac{1}{2} + m_s)^2. \tag{23}$$

The needed $\nu_c = eB/(2\pi m)$ (for a free electron in a magnetic field) is related to the observable eigenfrequencies by the Brown-Gabrielse invariance theorem [56],

$$(\nu_c)^2 = (\bar{\nu}_c)^2 + (\bar{\nu}_z)^2 + (\bar{\nu}_m)^2, \tag{24}$$

which applies despite the mentioned imperfection shifts of the three eigenfrequencies. The third term in Eq. 23, the leading relativistic correction [17] with $\delta/\nu_c \equiv h\nu_c/(mc^2) \approx 10^{-9}$, would add uncertainty to the measurement if cyclotron energy levels were not resolved.

The anomaly and spin-up cyclotron frequencies ($\bar{\nu}_a \approx 173$ MHz and $\bar{f}_c$ in Fig. 17b) are measured, since

$$\frac{g}{2} = \frac{\bar{\nu}_c + \bar{\nu}_a}{\nu_c} \simeq 1 + \frac{\bar{\nu}_a - \bar{\nu}_z^2/(2\bar{f}_c)}{\bar{f}_c + 3\delta/2 + \bar{\nu}_z^2/(2\bar{f}_c)}. \tag{25}$$

We use the approximation to the right which requires no measurement of $\bar{\nu}_m$. It incorporates an expansion of the invariance theorem [56], using $\bar{\nu}_c \gg \bar{\nu}_z \gg \bar{\nu}_m \gg \delta$. Corrections go as the product of $(\bar{\nu}_z/\bar{\nu}_c)^4 \sim 10^{-12}$ and a misalignment/harmonic distortion factor $\sim 10^{-4}$ [56].

A change in cyclotron or spin state is revealed by $\bar{\nu}_z$ shifts (Fig. 18a-b) of a one-electron self-exited oscillator (SEO) [6]. The electron's axial oscillation induces a signal in a resonant circuit that is amplified and fed back to drive the oscillation. QND couplings of spin and cyclotron energies to $\bar{\nu}_z$ [4] arise because saturated nickel rings (Fig 17a) produce a small magnetic bottle, $\Delta\mathbf{B} = \beta_2[(z^2 - \rho^2/2)\hat{\mathbf{z}} - z\rho\hat{\boldsymbol{\rho}}]$ with $\beta_2 = 1540$ T/m^2.

Anomaly transitions are induced by applying potentials oscillating at $\bar{\nu}_a$ to electrodes, to drive an off-resonance axial motion through the bottle's $z\boldsymbol{\rho}$ gradient. The electron sees the oscillating magnetic field perpendicular to $\mathbf{B}$ as needed to flip its spin, with a gradient that allows a simultaneous cyclotron transition. Cyclotron transitions are induced by microwaves with a transverse electric field that are injected into and filtered by the cavity. The electron samples the same magnetic gradient while $\bar{\nu}_a$ and $\bar{f}_c$ transitions are driven, because both drives are kept on, with one detuned slightly so that only the other causes transitions.

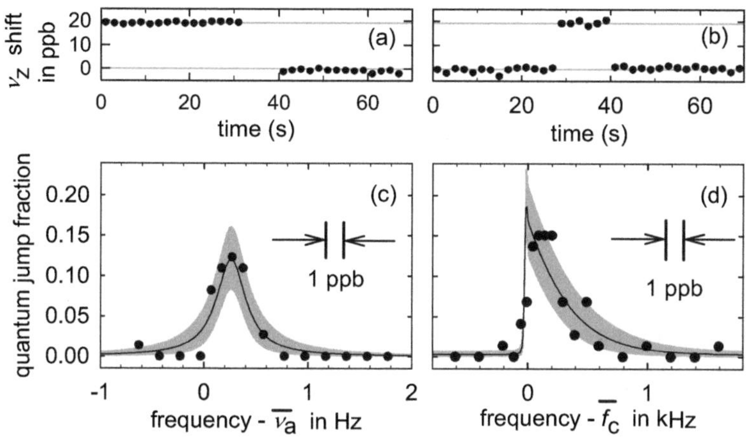

FIGURE 18. Sample $\bar{\nu}_z$ shifts for a spin flip (a) and for a one-quantum cyclotron excitation (b). Quantum jump spectroscopy lineshapes for anomaly (c) and cyclotron (d) transitions, with a maximum likelihood fit to the calculated lineshapes (solid). The bands indicate 68% confidence limits for distributions of measurements about the fit values.

A measurement starts with the SEO turned on to verify that the electron is in the upper of the two stable ground states, $|n = 0, m_s = 1/2\rangle$. Simultaneous $\bar{\nu}_c - \delta/2$ and $\bar{\nu}_a$ drives prepare this state as needed. The magnetron radius is reduced with 1.5 s of strong sideband cooling [17] at $\bar{\nu}_z + \bar{\nu}_m$, and the detection amplifier is turned off. After 1 s, either a $\bar{f}_c$ drive, or a $\bar{\nu}_a$ drive, is on for 2 s. The detection amplifier and the SEO are then switched on to check for a cyclotron excitation, or a spin flip (from an anomaly transition followed by a cyclotron decay). Inhibited spontaneous emission gives the time needed to observe a cyclotron excitation before an excited state decays. We step through each $\bar{\nu}_c$ and $\bar{\nu}_a$ drive frequency in turn, recording the number of quantum jumps per drive attempt. This measurement cycle is repeated during nighttimes, when electrical and magnetic noise are lower. A low drive strength keeps the transition probability below 20% to avoid saturation effects.

Quantum jump spectroscopy (measuring the quantum jumps per attempt to drive them as a function of drive frequency) gives resonance lineshapes for $\bar{f}_c$ and $\bar{\nu}_a$ (Fig. 18c-d). For weak drives that avoid saturation, the line shape comes from thermal axial motion within the magnetic bottle [45]. The small coherent axial oscillation at $\bar{\nu}_a$ has no noticeable effect. However, otherwise undetectable ppb fluctuations in **B**, on time scales shorter than an hour, would smear the expected lineshapes.

At the first of two magnetic fields used, $\bar{\nu}_c \approx 146.8$ GHz. A 1.4 s damping time gives good lineshape statistics (e.g., Fig. 18c-d) with 66 measurement cycles per night on average. Three methods to extract $\bar{\nu}_a$ and $\bar{f}_c$ from lineshapes give the same g within 0.6 ppt – our "lineshape model" uncertainty in Table 1. The first is maximum likelihood fitting of the Brownian motion lineshape. The second method fits a convolution of this lineshape and a Gaussian resolution function, about 1 ppb wide. The third method weights each drive frequency by the number of quantum jumps it produces, and uses the weighted average frequencies in Eq. 25 for $\bar{\nu}_a$ and $\bar{f}_c$. (Understood shifts proportional to axial temperature, common to both frequencies, do not increase the uncertainty.) This weighted average method should account for Brownian axial motion and additional fluctuations of **B**. At our second field, where $\bar{\nu}_c \approx 149.0$ GHz, the 6.7 s damping time allows only 29 measurement cycles per night on average. A long wait is needed to make certain that a spin flip has not occurred. The weighted averages method is used for the lower statistics lineshapes.

The $\bar{\nu}_z$ in Eq. 25 pertains while $\bar{f}_c$ and $\bar{\nu}_a$ are driven – not what is measured when the SEO amplifier is on and increasing the axial temperature from 0.3 to 5 K. Limits on axial heating shifts come from the width of a notch in the noise spectrum resonance for the resonant circuit [17] (Table 1), measured less well for $\bar{\nu}_c \approx 146.8$ GHz.

Source $\bar{\nu}_c$ =	**146.8 GHz**	**149.0 GHz**
$\bar{\nu}_z$ shift	0.2(0.3)	0.00(0.02)
Anomaly power	0.0(0.4)	0.00(0.14)
Cyclotron power	0.0(0.3)	0.00(0.12)
Cavity shift	12.8(5.1)	0.06(0.39)
Lineshape model	0.0 (0.6)	0.00 (0.60)
Statistics	0.0 (0.2)	0.00 (0.17)
Total (in ppt)	**13.0(5.2)**	**0.06(0.76)**

TABLE 1. Applied corrections and uncertainties for g in ppt.

Although the g-value from Eq. 25 is independent of **B**, field stability is still an important challenge, since $\bar{\nu}_a$ and $\bar{f}_c$ are measured at different times. After the superconducting solenoid settles for several months, field drifts below 10^{-9}/night have been observed. This requires regulating five He and N_2 pressures in the solenoid and experiment cryostats, and the surrounding air temperature to 0.3 K. We correct for drifts up to 10^{-9}/hr using a cyclotron resonance edge measured once in three hours.

The trap cavity modifies the density of states of radiation modes of free space, though not enough to significantly affect QED calculations of g [57]. However, cavity radiation modes do shift $\bar{f}_c$ [3] – still a significant uncertainty, as in the past [3, 18]. We use a synchronized-trapped-electrons method [15] to observe quantitatively understandable radiation modes (Fig. 19a) of a good cylindrical Penning trap cavity [1]. Our best measurement comes from choosing $\bar{\nu}_c \approx 149.0$ GHz, maximally detuned from modes that couple to a centered electron's cyclotron motion. A measurement at $\bar{\nu}_c \approx 146.8$ GHz, uncomfortably close to TE_{127}, checks how well cavity shifts are understood. Until the cavity spectrum and its frequency calibration is more carefully studied, TE_{127} and TM_{143} are assumed only to lie within the shaded bands. A renormalized calculation (Eq. 8.19 of [17]) gives a range of possible cavity shifts of the measured g (Fig. 19b) that is insensitive to mode quality factors for $Q > 500$. Assigned shifts and uncertainties are indicated in Fig. 19b and in Table 1. The first direct observation of a cavity shift of g, the difference between our two measurements (Fig. 19c), lies within the predicted range.

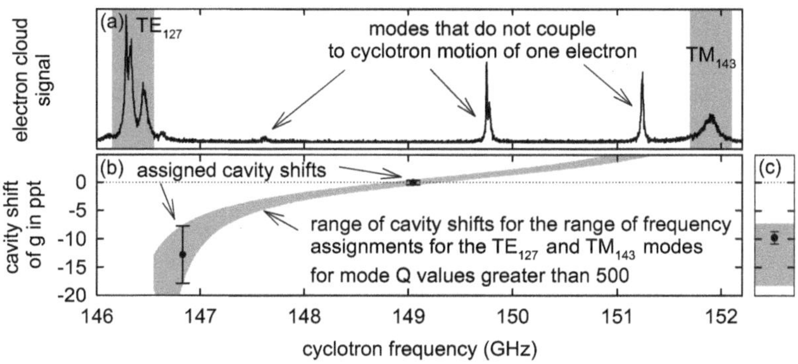

FIGURE 19. Modes of the trap cavity observed with synchronized electrons (a). Resulting assigned cavity shifts (points and Table 1) (b). First measured cavity shift of g (point) is the shift between measurements at 146.8 and 149.0 GHz (c). Gray bands are the assumed calibration and identification uncertainties for mode frequencies in (a), and the resulting range of predicted cavity shifts in (b) and (c).

A new value for the electron magnetic moment,

$$g/2 = 1.001\,159\,652\,180\,85\,(76)\ (0.76\ \text{ppt}), \tag{26}$$

comes from the measurement at $\bar{\nu}_c \approx 149.0$ GHz. (A weighted average with the more uncertain measurement at $\bar{\nu}_c \approx 146.8$ GHz is larger by 0.06 ppt, with a decreased uncertainty of 0.75 ppt.) The standard deviation, about six times smaller than from any previous measurement, arises mostly from the lineshape model and cavity shifts (Table 1). Varying the $\bar{\nu}_a$ and $\bar{f}_c$ drive power causes no detectable shifts of g.

QED provides an asymptotic series relating g and α,

$$\frac{g}{2} = 1 + C_2\left(\frac{\alpha}{\pi}\right) + C_4\left(\frac{\alpha}{\pi}\right)^2 + C_6\left(\frac{\alpha}{\pi}\right)^3 + C_8\left(\frac{\alpha}{\pi}\right)^4 + ... + a_{\mu\tau} + a_{\text{hadronic}} + a_{\text{weak}}, \quad (27)$$

with hadronic and weak contributions added, and assuming no electron substructure. Impressive calculations, summarized in the next section, give exact C_2, C_4 and C_6, a numerical value and uncertainty for C_8, and a small $a_{\mu\tau}$.

The next section details a new determination of α, from the measured g and Eq. 27,

$$\alpha^{-1}(\text{H06}) = 137.035\,999\,068\,(12)\,(30)\,(90) \quad (28)$$

$$= 137.035\,999\,068\,(96)\,[0.70\text{ ppb}], \quad (29)$$

The first line gives the experimental uncertainty first and the QED uncertainty second, including an estimated contribution from a yet uncalculated C_{10} [8]. The total 0.70 ppb uncertainty is ten times smaller than for the next most precise methods (Fig. 16b) – determining α from measured mass ratios, optical frequencies, together with either Rb [58] or Cs [59] recoil velocities.

The most stringent test of QED (one of the most demanding comparisons of any calculation and experiment) continues to come from comparing measured and calculated g-values, the latter using an independently measured α as an input. The next section shows that the new g, compared to Eq. 27 with α(Cs) or α(Rb), gives a difference $|\delta g/2| < 17 \times 10^{-12}$. The small uncertainties in $g/2$ will allow a ten times more demanding test if ever the large uncertainties in the independent α values can be reduced. The prototype of modern physics theories is thus tested far more stringently than its inventors ever envisioned [60], with better tests to come.

The same comparison of theory and experiment probes the internal structure of the electron [8, 51] – limiting the electron to constituents with a mass $m^* > m/\sqrt{\delta g/2} = 130$ GeV/c^2, corresponding to an electron radius $R < 1 \times 10^{-18}$ m. If this test was limited only by our experimental uncertainty in g, then we could set a limit $m^* > 600$ GeV. These high energy limits seem somewhat remarkable for an experiment carried out at 100 mK.

Are experimental improvements possible? A reduction of the 0.76 ppt uncertainty of the measured electron g seems likely, given that this fully-quantum measurement has only recently been realized. Time is needed to study the lineshapes and cavity shifts as a function of magnetic field, to improve cooling methods, and to make the magnetic field more stable.

In conclusion, greatly improved measurements of the electron magnetic moment and the fine structure constant, and a sensitive probe for internal electron structure, come from resolving the lowest cyclotron and spin levels of a one-electron quantum cyclotron. A self-excited oscillation of the electron reveals one-quantum transitions. A cylindrical Penning trap cavity narrows resonance lines by inhibiting spontaneous emission. Electromagnetic modes of this understandable cavity geometry, probed with synchronized electrons, shift g in a measurable way that can be corrected. The new $g/2$ differs from a long accepted value by 1.7 standard deviations, and its fractional uncertainty of 7.6×10^{-13} is nearly six times smaller. The new α has an uncertainty ten times smaller than that from any other method to determine the fine structure constant.

Measurement details and a preliminary analysis of the work in this section are in a thesis [61]. S. Peil, D. Enzer, and K. Abdullah contributed to earlier versions of the apparatus, and J. McArthur gave electronics support. The NSF AMO program provided long-term funding.

6. New Determination of the Fine Structure Constant

The new determination of the fine structure constant discussed in this section was initially reported in collaboration with my students D. Hanneke and B. Odom, and in collaboration with theorists T. Kinoshita and M. Nio [8]. This section has been updated to include the most recent correction to the QED evaluation [9].

The electron g-value, the dimensionless measure of the electron magnetic moment in terms of the Bohr magneton, provides our most sensitive probe into the structure of what is believed to be the only stable point particle with substantial mass – a particle that seems very insensitive to physics at small distance scales. The fundamental fine structure constant, $\alpha = e^2/(4\pi\epsilon_0\hbar c)$, gives the strength of the electromagnetic interaction, and is a crucial building block in our system of fundamental constants [54]. Quantum electrodynamics (QED), the wonderfully successful theory that describes the interaction of light and matter, provides an incredibly accurate prediction for the relationship between g and α, with only small corrections for short distance physics.

A new measurement of g [7] achieves an accuracy that is nearly six times more accurate than the last measurement of g back in 1987 [18]. An improved QED calculation that includes contributions from 891 Feynman diagrams [62] now predicts g in terms of α through order $(\alpha/\pi)^4$. Together, the newly measured g, with the more accurate QED calculation, determined a new and much more accurate determination of α. It is the first higher accuracy measurement of α since 1987 (Fig. 20), and is ten times more accurate than any other method to determine α. The most stringent test of QED is also presented.

Since $g = 2$ for a Dirac point particle, the dimensionless moment is often written as $g = 2(1 + a)$. The deviation a has come to be called the anomalous magnetic moment of the electron or sometimes simply the electron anomaly. It arises almost entirely from the vacuum fluctuations and polarizations that are described by QED,

$$a = a(QED) + a(hadron) + a(weak), \tag{30}$$

with only small additions for short distance physics, well understood in the context of the standard model [63] (unlike the case for the heavier muon lepton). Any additional contribution to the anomaly would therefore be extremely significant,

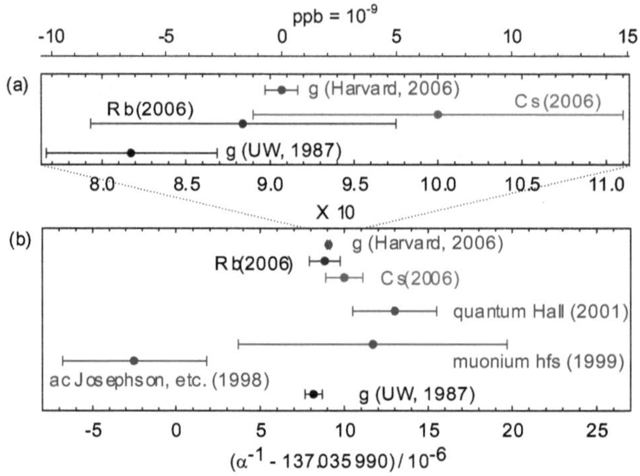

FIGURE 20. A ten times expanded scale (a) is requried to see the small uncertainties in the new α announced here. Traditional determinations can be seen on a larger scale (b). Measured g are converted to α using current QED theory.

indicating electron substructure [51], new short distance physics, or problems with QED theory (and perhaps with quantum field theory more generally).

A long tradition of improved measurements of g [52, 18] now continues after a hiatus of nearly twenty years. A new measurement achieves a much higher accuracy for g [7] by resolving the quantum cyclotron and spin levels of one electron suspended for months at a time in a cylindrical Penning trap. Quantum jump spectroscopy of transitions between these levels determines the spin and cyclotron frequencies, and $g/2$ is essentially the ratio of such measured frequencies. The cylindrical Penning cavity shapes the radiation field in which the electron is located, narrowing resonance linewidths by inhibiting spontaneous emission, and providing boundary conditions which make it possible to identify the symmetries of cavity radiation modes. A QND (quantum nondemolition) coupling, of the cyclotron and spin energies to the frequency of an orthogonal and nearly harmonic electron oscillation, reveals the quantum state. This harmonic oscillation of the electron is self-excited, by a signal derived from its own motion, to produce the large signal-to-noise ratio needed to quickly read out the quantum state without ambiguity.

The newly measured g has an uncertainty of only 7.6 parts in 10^{13}. Subtracting off the $g = 2$ for a Dirac point particle gives an anomaly [7]

$$a(\mathrm{H06}) = 1\,159\,652\,180.85\,(76) \times 10^{-12}\ [0.66\ \mathrm{ppb}]. \qquad (31)$$

As detailed in the previous section, the uncertainty mostly originates from three sources. The largest contribution arises from imperfect fits to the expected lineshape model; likely these can be understood and reduced with careful study. Tiny magnetic field instabilities are one possible cause. The second source of uncertainty is cavity shifts, caused when the cyclotron frequency of an electron in trap cavity is shifted by interactions with cavity radiation modes that are near in frequency. The frequencies of cavity radiation modes are measured well enough to identify the spatial symmetry of the modes, and to calculate and correct for cavity shifts to g from the known electromagnetic field configurations. A smaller third uncertainty is statistical, and could be reduced as needed with more measurements.

QED calculations involving many Feynman diagrams provide the coefficients for expansions in powers of the small ratio $\alpha/\pi \approx 2 \times 10^{-3}$. The QED anomaly

$$\begin{aligned} a(\mathrm{QED}) &= A_1 + A_2(m_e/m_\mu) + A_2(m_e/m_\tau) \\ &+ A_3(m_e/m_\mu, m_e/m_\tau), \end{aligned} \tag{32}$$

is a function of lepton mass ratios. Each A_i is a series,

$$A_i = A_i^{(2)}\left(\frac{\alpha}{\pi}\right) + A_i^{(4)}\left(\frac{\alpha}{\pi}\right)^2 + A_i^{(6)}\left(\frac{\alpha}{\pi}\right)^3 + \ldots. \tag{33}$$

The calculations are so elaborate that isolating and eliminating mistakes is a substantial challenge, as is determining and propagating numerical integration uncertainties.

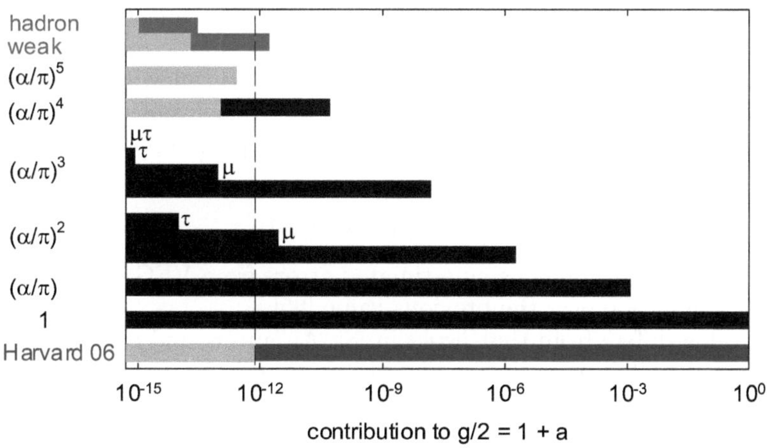

FIGURE 21. Contributions to $g/2$ for the experiment (green), terms in the QED series (black), and from small distance physics (blue). Uncertainties are in red. The μ, τ and $\mu\tau$ indicate terms dependent on mass ratios m_e/m_μ, m_e/m_τ and the two ratios, m_e/m_μ and m_e/m_τ, respectively.

Fig. 21 compares the contributions and uncertainties for $g/2$. The leading constants for second [64], third [65, 66, 67] and fourth [68, 69, 70, 71, 72] orders,

$$A_1^{(2)} = 0.5, \quad (34)$$

$$A_1^{(4)} = -0.328\,478\,965\,579\ \ldots, \quad (35)$$

$$A_1^{(6)} = 1.181\,241\,456\ \ldots, \quad (36)$$

have been evaluated exactly. The latter confirms the value 1.181 259 (4) obtained numerically [73]. Mass-dependent QED additions [74, 75, 76, 77, 78, 79],

$$\begin{aligned} A_2^{(4)}(m_e/m_\mu) &= 5.197\,386\,70\,(27) \times 10^{-7}, \\ A_2^{(4)}(m_e/m_\tau) &= 1.837\,63\,(60) \times 10^{-9}, \\ A_2^{(6)}(m_e/m_\mu) &= -7.373\,941\,58\,(28) \times 10^{-6}, \\ A_2^{(6)}(m_e/m_\tau) &= -6.581\,9\,(19) \times 10^{-8}, \\ A_3^{(6)}(m_e/m_\mu, m_e/m_\tau) & \\ &= 0.190\,95\,(63) \times 10^{-12}, \end{aligned} \quad (37)$$

make only very small contributions. Uncertainties derive from the uncertainties in measured lepton mass ratios.

Crucial progress came in evaluating, checking, and determining the uncertainty in the eighth order $A_1^{(8)}$, which includes contributions of 891 Feynman diagrams. Typical diagrams of the 13 gauge invariant subgroups are shown in Fig. 22. Integrals of 373 of these (Groups I - IV) have been verified (and corrected) by more than one independent formulation [80, 62]. The 518 diagrams of Group V, with no closed lepton loops, await completion of an independent verification. However, their renormalization terms are derived by systematic reduction of original integrands applying a simple power-counting rule [81], allowing extensive cross-checking among themselves and with exactly known diagrams of lower order [82]. Numerical integrations with VEGAS [83], on many supercomputers over more than 10 years [62, 84], then yields

$$A_1^{(8)} = -1.9144\,(35) \quad (38)$$

The uncertainty, determined using estimated errors from VEGAS, is improved by an order of magnitude over the previous value [85].

This value differs from that listed in [8]. An automated code generator [62], produced to calculate the tenth-order contribution to $g/2$, was used to examine the 518 of 891 eighth-order QED diagrams that had no previous independent check. Only 47 integrals represent the 518 vertex diagrams when the Ward-Takahashi identity and time-reversal invariance are used. A diagram-by-diagram comparison with the previous calculation [86] showed that 2 of the 47 require a corrected treatment of infrared divergences [84], leading to the corrected value used here.

Is it likely that other adjustments of the QED theory will shift the α that is determined from the electron g? Hopefully not, now that all eighth-order contributions have been checked independently by two or more methods for the first time. What could further shift this determination of α would be a larger-than-expected tenth-order QED contribution to g/2 – now being evaluated using the new computational method that revealed the need for this update.

The summary of precise α determinations (Fig. 20) differs from that of one year earlier [8] in several ways. The corrected QED evaluation shifts the α from the Harvard and UW g measurements. The atom-recoil determination of α(Rb) shifts due to an experimental correction [87]. The neutron α is no longer included awaiting a change required by the reevaluations of the Si lattice constant and its uncertainties (eg. [88]).

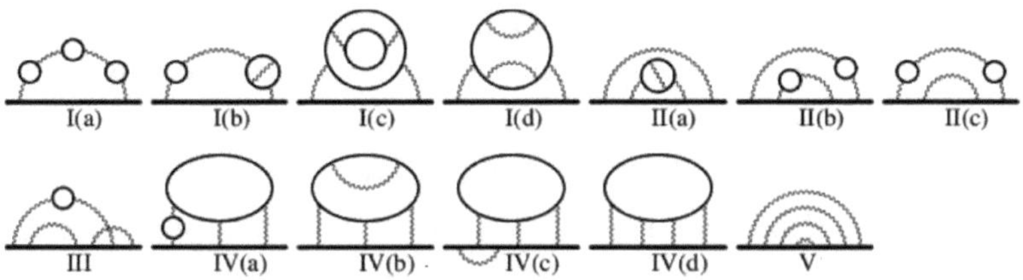

FIGURE 22. Typical diagrams from each gauge invariant subgroup that contibutes to the eighth-order electron magnetic moment. Solid and wiggly curves represent the electron and photon, respectively. Solid horizontal lines represent the electron in an external magnetic field.

The high experimental precision makes the tenth order contribution to g potentially important if the unknown $A_1^{(10)}$ is unexpectedly large, though this seems unlikely. To get a feeling for its possible impact we use a bound

$$|A_1^{(10)}| < x \tag{39}$$

with an estimate $x = 3.8$ [54], while awaiting a daunting evaluation of contributions from 12672 Feynman diagrams that is now underway [62].

Also owing the high precision, non-QED contributions,

$$\begin{aligned} a(\text{hadron}) &= 1.671\ (19) \times 10^{-12}, \\ a(\text{weak}) &= 0.030\ (01) \times 10^{-12}, \end{aligned} \tag{40}$$

must be included. Fortunately, these are small and well understood in the context of the standard model [54, 63]. They are much larger and more important for the muon.

The new experimental measurement of g, and the greatly improved QED calculation, thus determine a value of α that is ten times more accurate than any other method to determine α,

$$\alpha^{-1}(\text{H06}) = 137.035\,999\,068\ (12)\ (30)\ (90) \tag{41}$$

$$= 137.035\,999\,068\ (96)\ [0.70\ \text{ppb}], \tag{42}$$

In the first line, the first uncertainty is from the calculated $A_1^{(8)}$ and the last is from the measured g. The middle uncertainty is from the estimated bound on the unknown $A_1^{(10)}$ in Eq. (39). More generally it is $(8x)$, which rounds to (30) for the estimate $x = 3.8$ [54].

Note that while the tenth order impact on α is not large since $8x \ll 90$, a real calculation will be needed before a much more accurate α can be deduced from a better g. Note also that the exact $A_1^{(6)}$ of Laporta and Remiddi, in Eq. (36) and [72], eliminates an earlier numerical uncertainty [73] that would add (60) to the list of three uncertainties in Eq. (41), significantly increasing the total uncertainty in Eq. (42).

Testing QED requires an independent measurement of α, to be used with QED theory, to determine an anomaly that can be compared to what is measured. Recent measurements utilizing Cs and Rb atoms yield

$$\alpha^{-1}(Cs06) = 137.036\,000\,00\,(110) \quad [8.0\text{ ppb}], \tag{43}$$

$$\alpha^{-1}(Rb06) = 137.035\,998\,84\,(\;91) \quad [6.6\text{ ppb}]. \tag{44}$$

The latest versions rely upon many experiments, including the measured Rydberg constant [89], the Cs or Rb mass in amu [90], and the electron mass in amu [91, 92]. The needed $\hbar/M[Cs]$ comes from an optical measurement of the Cs D1 line [59, 93], and the "preliminary" recoil shift for a Cs atom in an atom interferometer [94]. The needed $\hbar/M[Rb]$ come from a measurement of an atom recoil of a Rb atom in an optical lattice [58] as recently corrected[87].

The Cs and Rb determinations of α, together with QED theory (and hadronic and weak corrections), give

$$a(Cs06) = 1\,159\,652\,173.00\,(0.10)\,(0.31)\,(9.30) \times 10^{-12},$$

$$a(Rb06) = 1\,159\,652\,182.80\,(0.10)\,(0.31)\,(7.70) \times 10^{-12}.$$

Uncertainties are from the eighth order calculation, the estimated tenth order limit, and the determinations of α. Calculated and measured anomalies differ by

$$a(\text{Cs06}) - a(\text{H06}) = -\;7.9\;(9.3) \times 10^{-12}, \tag{45}$$

$$a(\text{Rb06}) - a(\text{H06}) = 1.9\;(7.7) \times 10^{-12}, \tag{46}$$

with the uncertainties limited by the uncertainties in $\alpha(Cs)$ and $\alpha(Rb)$, in Eqs. (43)-(44).

What theory improvements might be expected in the future? The theory contribution to the uncertainty in the new α is less than that from experiment by a factor of 3. The eighth order uncertainty in $A_1^{(8)}$ can be reduced with the accumulation of better statistics in the numerical evaluation of integrals. Ambitious efforts underway aim for an analytic evaluation of this coefficient [95]. Another big theory challenge is in evaluating the tenth-order coefficient, $A_1^{(10)}$, with the mentioned contributions from 12672 Feynman diagrams. Work underway suggests that it should be possible to evaluate $A_1^{(10)}$ to a few percent, reducing the theoretical uncertainty in a_e to 0.1 ppb or less.

What experimental improvements can be expected? A reduction of the 0.76 ppt uncertainty of the measured electron g seems likely, given that this fully quantum measurement has only recently been realized. With time to study the lineshapes and cavity shifts at many values of the magnetic field, to improve cooling methods, and to incorporate a more stable magnetic field, an improved α from g and QED seems quite possible. Experiments are also underway under the assumption that a substantially higher accuracy can be achieved in atom recoil measurements – currently the weak link in determinations of α that are independent of g and QED. A ten-fold improved accuracy in the independent α would allow a QED test that is more stringent than current tests by this factor, even without improved measurement of g, or more accurate QED theory.

In conclusion, a slightly shifted and much more accurate determination of the fine structure constant comes from the new measurement of the electron g value, and improved QED theory. It is more accurate than any other method to determine α by a factor of ten. The working assumption is that the electron has no internal structure that is not described by QED, nor by small distance scale physics other than what is well understood within the context of the standard model of particle physics. Comparing the α from g and QED, to the α determined with Cs and Rb atoms, shows that QED continues to be a superb description of the interaction of atoms and light. However, a ten times more stringent QED test awaits a determination of α that is independent of the electron g and QED, but achieves the accuracy in the α reported here. We expect more accurate measurements of the electron g, along with more accurate QED calculations, and are thus optimistic that more accurate future determinations of α may be possible.

Experiments at Harvard to measure g used in this section were supported by the NSF AMO experimental program. Theory work by T.K. was supported by the NSF theory program of the US, the Eminent Scientist Invitation Program of RIKEN, Japan, and a grant-in-aid from Japan's Ministry of Education, Science and Culture. M.N. was partly supported by a JSPS grant-in-aid, and used computational resources of the RIKEN Super Combined Cluster System.

7. Almost Outdated

As these notes are being prepared for publication, my student D. Hanneke and I are nearly finished with our analysis of new measurements of the electron magnetic moment and the fine structure constant that look to have a precision that is two to three time better than what is reported here. Stay tuned.

8. Conclusion

Quantum jump spectroscopy of the lowest spin and cyclotron energy levels of a one-electron quantum cyclotron yields the most precise measurement of the electron magnetic moment, and the most precise measurement of the fine structure constant. The electron is suspended for months at a time in a cylindrical Penning trap, the electrodes of which provide the electrostatic quadrupole potential for trapping the electron, and also modify the radiation field and density of states as needed to inhibit the spontaneous emission of synchrotron radiation. Feedback methods provide cooling possibilities and turn the electron into a one-particle self-excited oscillator that is used for QND measurements of the cyclotron and spin states. One spin-off measurement being pursued, in collaboration with my student S. Fogwell, is making the most stringent test of CPT invariance with a lepton system, by comparing measured g values for the electron and positron. Another spin-off measurement underway, in collaboration with my student N. Guise, is to make a one-antiproton self-excited oscillator to measure the antiprotons magnetic moment a million times more accurately than has been possible so far. A third spin-off measurement being contemplated is a direct measurement of the electron-to-proton mass ratio.

References

[1] G. Gabrielse and F. C. MacKintosh, Intl. J. of Mass Spec. and Ion Proc. **57**, 1 (1984).

[2] J. N. Tan and G. Gabrielse, Appl. Phys. Lett. **55**, 2144 (1989).

[3] L. S. Brown, G. Gabrielse, K. Helmerson, and J. Tan, Phys. Rev. Lett. **55**, 44 (1985).

[4] S. Peil and G. Gabrielse, Phys. Rev. Lett. **83**, 1287 (1999).

[5] B. D'Urso, B. Odom, and G. Gabrielse, Phys. Rev. Lett. **90**, 43001 (2003).

[6] B. D'Urso, R. Van Handel, B. Odom, D. Hanneke, and G. Gabrielse, Phys. Rev. Lett. **94**, 113002 (2005).

[7] B. Odom, D. Hanneke, B. D'Urso, and G. Gabrielse, Phys. Rev. Lett. **97**, 030801 (2006).

[8] G. Gabrielse, D. Hanneke, T. Kinoshita, M. Nio, and B. Odom, Phys. Rev. Lett. **97**, 030802 (2006).

[9] G. Gabrielse, D. Hanneke, T. Kinoshita, M. Nio, and B. Odom, Phys. Rev. Lett. **99**, 039902 (2007).

[10] X. Maître, E. Hagley, G. Nogues, C. Wunderlich, P. Goy, M. Brune, J. M. Raimond, and S. Haroche, Phys. Rev. Lett. **79**, 769 (1997).

[11] M. Weidinger, B. T. H. Varcoe, R. Heerlein, and H. Walther, in *Abstracts of ICAP 16* (Univ. of Windsor, Windsor, 1998), p. 362.

[12] P. Domokos, M. Brune, J. M. Raimond, and S. Haroche, epjd **1**, 1 (1998).

[13] D. M. Meekhof, C. Monroe, B. E. King, W. M. Itano, and D. J. Wineland, Phys. Rev. Lett. **76**, 1796 (1996).

[14] I. Bouchoule, H. Perrin, A. Kuhn, M. Morinaga, , and C. Salomon, Phys. Rev. A **59**, R8 (1999).

[15] J. Tan and G. Gabrielse, Phys. Rev. Lett. **67**, 3090 (1991).

[16] G. Gabrielse and J. N. Tan, *"One Electron in a Cavity"* (Academic Press, New York, 1994), pp. 267–299.

[17] L. S. Brown and G. Gabrielse, Rev. Mod. Phys. **58**, 233 (1986).

[18] R. S. Van Dyck, Jr., P. B. Schwinberg, and H. G. Dehmelt, Phys. Rev. Lett. **59**, 26 (1987).

[19] K. S. Thorne, R. W. P. Drever, and C. M. Caves, Phys. Rev. Lett. **40**, 667 (1978).

[20] V. B. Braginsky and F. Y. Khalili, Rev. Mod. Phys. **68**, 1 (1996).

[21] M. Rigo, G. Alber, F. Mota-Furtado, and P. F. O'Mahony, Phys. Rev. A **55**, 1665 (1997).

[22] G. Gabrielse and H. Dehmelt, Phys. Rev. Lett. **55**, 67 (1985).

[23] M. Strutt and A. V. der Ziel, Physica **IX**, 513 (1942).

[24] C. Kittel, '*"Elementary Statistical Physics"* (Wiley, New York, 1958), pp. 141–156.

[25] B. B. V. I. J. M. W. Milatz, J. J. Van Zolingen, Physica **19**, 195 (1953).

[26] P. G. Roll, R. Krotkov, and R. H. Dicke, Ann. of Phys. **26**, 442 (1964).

[27] R. L. Forward, J. Appl. Phys. **50**, 1 (1979).

[28] B. E. Bernard and R. C. Ritter, J. Appl. Phys. **64**, 2833 (1988).

[29] P. F. Cohadon, A. Heidmann, and M. Pinard, Phys. Rev. Lett. **83**, 3174 (1999).

[30] S. van der Meer, Rev. Mod. Phys. **57**, 689 (1985).

[31] H. Dehmelt, W. Nagourney, and J. Sandberg, Proc. Natl. Acad. Sci. USA **83**, 5761 (1986).

[32] S. L. Rolston and G. Gabrielse, hyper **44**, 233 (1988).

[33] N. Beverini, V. Lagomarsino, G. Manuzio, F. Scuri, G. Testera, and G. Torelli, Phys. Rev. A **38**, 107 (1988).

[34] J.-M. Courty, F. Grassia, and S. Reynaud, arXiv:quant-ph/0110021 (2001).

[35] J. B. Johnson, Phys. Rev. **32**, 97 (1928).

[36] H. Nyquist, Phys. Rev. **32**, 110 (1928).

[37] To be published.

[38] S. Rainville, M. Bradley, J. Porto, J. Thompson, and D. Pritchard, Hyperfine Int. **132**, 177 (2001).

[39] A. Bain, *patent no. 8783*, Great Britain (1841).

[40] T. R. Albrecht, P. Grütter, D. Horne, and D. Rugar, J. Appl. Phys. **69**, 668 (1991).

[41] A. N. Cleland and M. L. Roukes, Nature **392**, 160 (1998).

[42] M. Lax, Phys. Rev. **160**, 290 (1967).

[43] W. B. Davenport, Jr., J. Appl. Phys. **24**, 720 (1953).

[44] W. F. Walker, T. K. Sarkar, F. I. Tseng, and D. D. Weiner, IEEE Trans. Inst. Meas. **31**, 239 (1982).

[45] L. S. Brown, Ann. Phys. (N.Y.) **159**, 62 (1985).

[46] R. S. Van Dyck, Jr., P. B. Schwinberg, and H. G. Dehmelt, Phys. Rev. Lett. **38**, 310 (1977).

[47] W. Quint and G. Gabrielse, Hyperfine Interact. **76**, 379 (1993).

[48] S. Eidelman, et al., Phys. Lett. B **592**, 1 (2004).

[49] G. Gabrielse, A. Khabbaz, D. S. Hall, C. Heimann, H. Kalinowsky, and W. Jhe, Phys. Rev. Lett. **82**, 3198 (1999).

[50] H. Häffner, T. Beier, N. Hermanspahn, H.-J. Kluge, W. Quint, S. Stahl, J. Verdú, and G. Werth, Phys. Rev. Lett. **85**, 5308 (2000).
[51] S. J. Brodsky and S. D. Drell, Phys. Rev. D **22**, 2236 (1980).
[52] A. Rich and J. C. Wesley, Rev. Mod. Phys. **44**, 250 (1972).
[53] R. S. Van Dyck Jr. , P. B. Schwinberg, and H. G. Dehmelt, *The Electron* (Kluwer Academic Publishers, Netherlands, 1991).
[54] P. J. Mohr and B. N. Taylor, Rev. Mod. Phys. **77**, 1 (2005).
[55] G. W. Bennett and et al., Phys. Rev. D **73**, 072003 (2006).
[56] L. S. Brown and G. Gabrielse, Phys. Rev. A **25**, 2423 (1982).
[57] D. G. Boulware, L. S. Brown, and T. Lee, Phys. Rev. D **32**, 729 (1985).
[58] P. Cladé, E. de Mirandes, M. Cadoret, S. Guellati-Khélifa, C. Schwob, F. Nez, L. Julien, and F. Biraben, Phys. Rev. Lett. **96**, 033001 (2006a).
[59] V. Gerginov, K. Calkins, C. E. Tanner, J. McFerran, S. Diddams, A. Bartels, and L. Hollberg, Phys. Rev. A **73**, 032504 (2006).
[60] F. Dyson, private communication.
[61] B. Odom, Ph.D. thesis, Harvard University (2004).
[62] T. Aoyama, M. Hayakawa, T. Kinoshita, and M. Nio, Nucl. Phys. **B740**, 138 (2006a).
[63] A. Czarnecki, B. Krause, and W. J. Marciano, Phys. Rev. Lett. **76**, 3267 (1996).
[64] J. Schwinger, Phys. Rev. **73**, 416L (1948).
[65] C. M. Sommerfield, Phys. Rev. **107**, 328 (1957).
[66] C. M. Sommerfield, Ann. Phys. (N.Y.) **5**, 26 (1958).
[67] A. Petermann, Helv. Phys. Acta **30**, 407 (1957).
[68] S. Laporta and E. Remiddi, Phys. Lett. B **265**, 182 (1991).
[69] S. Laporta, Phys. Rev. D **47**, 4793 (1993a).
[70] S. Laporta and E. Remiddi, Phys. Lett. B **356**, 390 (1995).
[71] S. Laporta, Phys. Lett. B **343**, 421 (1995).
[72] S. Laporta and E. Remiddi, Phys. Lett. B **379**, 283 (1996).
[73] T. Kinoshita, Phys. Rev. Lett. **75**, 4728 (1995).
[74] M. A. Samuel and G. Li, Phys. Rev. D **44**, 3935 (1991).
[75] G. Li, R. Mendel, and M. A. Samuel, Phys. Rev. D **47**, 1723 (1993).
[76] A. Czarnecki and M.Skrzypek, Phys. Lett. B **449**, 354 (1999).
[77] S. Laporta, Nuovo Cim. A **106A**, 675 (1993b).
[78] S. Laporta and E. Remiddi, Phys. Lett. B **301**, 440 (1993).
[79] B. Lautrup, Phys. Lett. **69B**, 109 (1977).
[80] T. Kinoshita and M. Nio, Phys. Rev. Lett. **90**, 021803 (2003).
[81] P. Cvitanovic and T. Kinoshita, Phys. Rev. D **10**, 4007 (1974).
[82] T. Kinoshita, *Theory of the Anomalous Magnetic Moment of the Electron – Numerical Approach* (World Scientific, Singapore, 1990).
[83] G. P. Lepage, J. Comput. Phys. **27**, 192 (1978).
[84] T. Aoyama, M. Hayakawa, T. Kinoshita, and M. Nio, hep-ph/0706.3496 (2007).
[85] V. W. Hughes and T. Kinoshita, Rev. Mod. Phys. **71**, S133 (1999).

[86] T. Aoyama, M. Hayakawa, T. Kinoshita, and M. Nio, Phys. Rev. D **73**, 013003 (2006b).
[87] P. Cladé, E. de Mirandes, M. Cadoret, S. Guellati-Khélifa, C. Schwob, F. Nez, L. Julien, and F. Biraben, Phys. Rev. A **74**, 052109 (2006b).
[88] P. Becker, G. Cavagnero, U. Kuetgens, G. Mana, and E. Massa, IEEE T. on Instrum. Meas. **56**, 230 (2007).
[89] C. Schwob, L. Jozefowski, B. de Beauvoir, L. Hilico, F. Nez, L. Julien, F. Biraben, O. Acef, J. J. Zondy, and A. Clairon, Phys. Rev. Lett. **82**, 4960 (1999).
[90] M. P. Bradley, J. V. Porto, S. Rainville, J. K. Thompson, and D. E. Pritchard, Phys. Rev. Lett. **83**, 4510 (1999).
[91] T. Beier, H. Häffner, N. Hermanspahn, S. G. Karshenboim, H.-J. Kluge, W. Quint, S. Stahl, J. Verdú, and G. Werth, Phys. Rev. Lett. **88**, 011603 (2002).
[92] D. L. Farnham, R. S. Van Dyck, Jr., and P. B. Schwinberg, Phys. Rev. Lett. **75**, 3598 (1995).
[93] T. Udem, J. Reichert, R. Holzwarth, and T. W. Hänsch, Phys. Rev. Lett. **82**, 3568 (1999).
[94] A. Wicht, J. M. Hensley, E. Sarajlic, and S. Chu, Phys. Scr. **T102**, 82 (2002).
[95] S. Laporta and E. Remiddi, private communication.

Gerald Gabrielse
Leverett Professor of Physics
Harvard University
Cambridge, MA 02138, USA
e-mail: gabrielse@physics.harvard.edu

Note added by the Editors

Since this paper was written, Professor Gabrielse and his group have obtained and published even more precise results on the electron magnetic moment and the fine structure constant [96]:

$$g/2 = 1.00115965218073(28) \quad [0.28ppt] \tag{47}$$

with an uncertainty 2.7 and 15 times smaller than for previous measurements in 2006 and 1987, and

$$\alpha^{-1} = 137.035999084(51) \quad [0.37ppb] \tag{48}$$

with an uncertainty 20 times smaller than for any independent determination of α.

[96] D. Hanneke, S. Fogwell, and G. Gabrielse, New Measurement of the Electron Magnetic Moment and the Fine Structure Constant, Phys. Rev. Lett. **100**, 120801 (2008).

The Spin
B. Duplantier, J.-M. Raimond and V. Rivasseau, Eds., 147–157

Poincaré Seminar 2007

The 2007 Nobel Prize in Physics: Albert Fert and Peter Grünberg

Vincent Cros, Albert Fert, Pierre Sénéor and Frédéric Petroff

Albert Fert and **Peter Grünberg** are well-known for having opened a new route in science and technology by their **discovery of the Giant MagnetoResistance (GMR)** in 1988. Soon after this discovery, the exploitation of the GMR had a considerable impact in **data storage technologies**. The GMR-based read heads developed by several major companies have led to a considerable increase of the density of information stored on computer's hard discs drives (HDD) of today and are the most important of the multiple applications of the GMR. The discovery of GMR kicked off the development of a new field in science and technology called **spintronics** in considerable expansion today. Not only Albert Fert and Peter Grünberg are regarded as the fathers of spintronics but they have also made outstanding contributions to its development until today. Their recent work on magnetic tunnel junctions, current-induced magnetization switching, microwave generation by spin transfer, or applications of carbon nanotubes in spintronics will certainly have an important impact on the technology of the near future.

1. The roots of spintronics and the discovery of GMR

GMR and spintronics take their roots in the pioneering work of Albert Fert around 1970 on the influence of the spin on the mobility of electrons in ferromagnetic materials [1]. After having experimentally demonstrated that, in a ferromagnetic metal, the electrons of opposite spin directions (spin up and spin down along the magnetization axis) carry different currents (as originally suggested by Mott), Fert worked out the well known two current model of the electrical conduction in ferromagnetic metals. He also showed that very large spin asymmetries of the conduction can be obtained by doping the ferromagnetic metal with impurities selected to scatter very differently the spin up and spin down electrons (iron or cobalt impurities in nickel, for example, scatter the spin down electrons 20 times more strongly than the spin up electrons). Moreover, some experiments of Fert on ternary alloys were already introducing the idea that he will exploit later to produce the GMR effects. He showed that the resistivity of a ternary alloy, for example $N_{1-x}(A_{x-y}, B_y)$, is strongly enhanced if the scattering by the impurities A and B have inverse spin asymmetries. Replacing the impurities A and B by magnetic layers A and B, one equally expects a large enhancement of the resistivity when their magnetizations are in opposite directions, which the basic concept of the GMR. However, this concept can work only if the thickness of the layers is in the nanometer range. The fabrication of multilayers with thicknesses in this range became technologically possible in the mid-eighties and, in particular, the growth of magnetic multilayers by Molecular Beam Epitaxy (MBE) was developed in the groups of Albert Fert and Peter Grünberg.

Another important step, two years before the discovery of the GMR, came from the demonstration by Brillouin scattering experiments in the group of Peter Grünberg that the magnetizations of two layers of iron separated by an ultra-thin layer of chromium were spontaneously oriented in opposite directions by an antiferromagnetic exchange interaction [2]. This was opening the way to the design of Fe/Cr multilayered structures in which the magnetizations of the adjacent magnetic layers can be switched from their spontaneous opposite orientations to parallel by applying a magnetic field. The GMR was discovered almost simultaneously in 1988 on Fe/Cr multilayers by Albert Fert [3] and on $Fe/Cr/Fe$ trilayers by Peter Grünberg [4]. Fert's paper presented both the experimental results and their interpretation on the basis of his previous work on the spin dependent conduction in ferromagnetic materials. As the change of resistance between the parallel and antiparallel magnetic configurations of the Fe/Cr multilayer was as large as 80%, Fert also coined the expression "Giant Magnetoresistance" to describe such huge effects (see Figure 1). The paper of Grünberg reporting the results on the Fe/Cr/Fe trilayers in 1989 was preceded by an outstanding patent filed in 1988, introducing the concept of Spin Valve (SV), and anticipating most of the present applications of the GMR (US patent 4949039, 1990). The discovery of the GMR created a considerable stir, first because it immediately turned out that it was opening a new field of research (called today spintronics), and secondly because the potential of applications became rapidly very clear as detailed in the final section. In 2003, with 2455 citations (more than 3500 in 2007), Fert's article of 1988 was ranked 6 in the Top Ten of the most cited Physical Review Letters since the creation of the review in 1953.

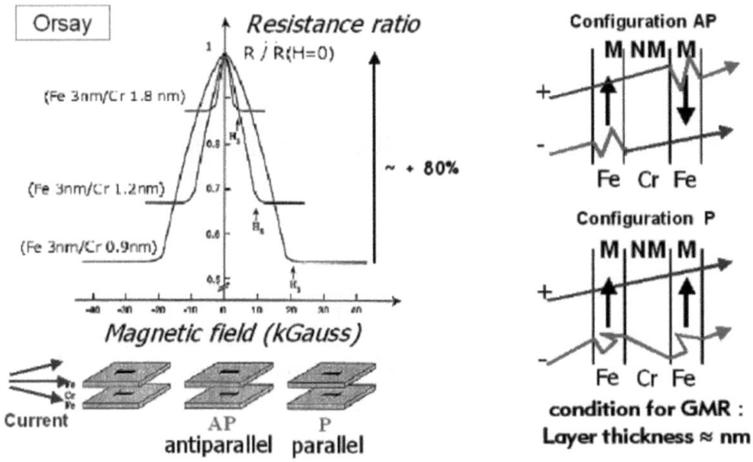

FIGURE 1. (left) Variation of the resistance as a function of the applied magnetique field (in kG) for an Fe/Cr multilayer.The decrease of the resistance that occurs when the magnetic field is applied is related to the change from an antiparallel configuration of the two magnetizations to a parallel configuration at high field. That is the Giant Magnetoresistance effect (GMR). (right) Illustration of the mechanism of GMR in terms of spin dependent scattering processes for the two configurations of the magnetizations i.e. antiparallel or parallel.

2. From GMR to spintronics

The goal of this section is not to review the entire field of spintronics but rather highlight some major steps and the contributions of Fert and Grünberg. The discovery of GMR triggered immediately an extensive research on magnetic multilayers and GMR attracting researchers worldwide in the new field of spintronics. On the experimental side, it was rapidly shown that GMR effects could be obtained not only with samples grown by molecular beam epitaxy but also with the faster and cheaper method of sputtering. In 1991, the GMR of the Co/Cu system was simultaneously discovered on samples grown by sputtering in Fert's group [5] and S. Parkin's group [6]. Co/Cu became the archetypal structure for GMR and inspired the advanced materials used today in the sensors of HDD's read heads. On the theory side, the first semi-classical model of the GMR was published in 1989 Phys. Rev. Lett. by Camley and Barnas in the laboratory of Peter Grünberg. Together with Levy and Zhang, Fert equally worked out the first quantum mechanical theory of the GMR in 1990 [7]. Initially studied in the CIP configuration (Current In the Plane), experiments soon followed exploring the Current Perpendicular to the Plane (CPP) geometry. CPP-GMR is not only interesting for applications but also because it has revealed spin accumulations effects which were analyzed in a seminal paper [8]. These effects play a major role in the most recent developments of spintronics such as spintronics with semiconductors [9], or molecular spintronics [10] which will not be detailed here.

The second important step in the history of spintronics is related to the introduction of the Magnetic Tunnel Junctions (MTJ). A MTJ is another type of SV in which the two magnetic layers are separated by an ultra-thin insulating layer (around 1 nm) instead of a metallic layer. Its electrical resistance is different for the parallel and antiparallel magnetic configurations of the electrodes, what is called the Tunneling MagnetoResistance effect (TMR). Some early observations of TMR effects, small and at low temperature, were already reported in 1975, but they had been hardly reproducible during 20 years. It is only in 1995 that large (20% for the ratio of the resistance change to the resistance of the parallel state) and reproducible effects were obtained by Moodera (MIT) and Miyazaki (Sendai) groups on MTJ with a tunnel barrier of amorphous alumina. Among others, Fert and Grünberg have actively contributed to the research on TMR. Grünberg studied the potential of semiconducting tunnel barriers [11]. On the fundamental side, a better understanding of the physics of MTJ came from the classical paper in which Fert demonstrated the active role of the electronic properties of the barrier on the TMR as shown in Figure 2 (Science 286, 507, 1999). Also new directions were explored to obtain much higher TMR than with the alumina-based MTJ. Very high TMR ratios (1800% at low temperature) were obtained in the group of Fert with MTJ based on the magnetic oxide $La_{2/3}Sr_{1/3}MnO_3$ see Figure 3 [12]. The exploration of these new directions was of high fundamental interest, but, for applications, the $La_{2/3}Sr_{1/3}MnO_3$-based MTJ had a too low TMR at room temperature. The real breakthrough for the TMR came from the development of MgO-based single crystal MTJ. The first publication reporting TMR effects on MgO-based MTJ was published by the group of Fert in collaboration with a Spanish group and showed that, with similar electrodes, the TMR could be slightly larger with MgO than with alumina [13]. This triggered the research on similar junctions in several groups, and, in 2004, the major breakthrough was the demonstration by an IBM group and by a Japanese group that the TMR ratio of MgO-based MTJ could reach 200% at room temperature. It was the kick off of an intense research on MgO-based MTJ which has led to TMR ratios as high as 500% together with relatively small resistances needed for many applications. These remarkable MTJ will have multiple applications (see last section). From a technological point of view, the interest of the MTJ with respect to the metallic spin valves comes from the vertical direction of the current and from the resulting possibility of a reduction of the lateral size to a submicronic scale by lithographic techniques. As a third step, the growth of semiconductor heterostructures incorporating ferromagnetic material is a challenge for today's spintronic. The general objective is to combine spintronic and traditional electronic. Semiconductor materials bring also long spin life time, well define energy levels when confine to $0D$ $1D$ or $2D$ and coupling with optical properties which make them a unique tool

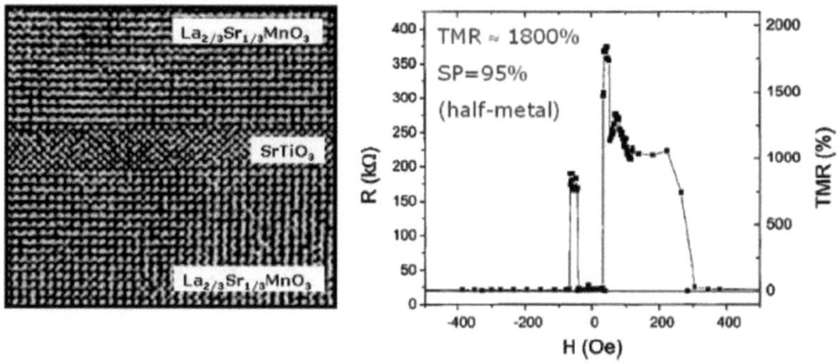

FIGURE 2. (left) High resolution TEM image of a $La_{2/3}Sr_{1/3}MnO_3/$ $SrTiO_3/$ $La_{2/3}Sr_{1/3}MnO_3$ heterostructure deposited by pulsed laser ablation. (right) Resistance as function of the field for a similar heterostructure measured at 4.2 K. The TMR ratio, about 1800% corresponds to a very large spin polarization (95%) of this magnetic oxyde.

to explore new spin related effect [14]. Classically 2 roads are follow, the first one take advantage of the properties of well know transition metal whereas the second investigate diluted magnetic semiconductors as $GaMnAs$. The recent theoretical advances lead to integration of well controlled interfacial resistance and demonstration of efficient spin injection. As an example we recently observed more than 50% of circular light polarization when injecting from $CoFeB/MgO$ into a III-V light emitting diode. The next step is the fabrication of a complete electrical device.
In the case of $GaMnAs$ related junctions, the complexity of the transport mechanisms associated with spin-orbit coupled states make this material a powerful means for finding novel effects and provides new challenges for theoretical understandings. This includes tunnel magnetoresistance (TMR) across single and double barriers [15], Spin transfer torque effect (STT) [16] and tunnel anisotropic magnetoresistance (TAMR). As an illustration (see Figure 3), the resonant TAMR on a GaAs quantum well can be used as a probe of the GAMnAs valence band anisotropy [17]. Nanospintronics is another prospective field at the interface between spintronics and nanophysics and deals with spin injection and detection in objects with reduced dimensions such as wires ($1D$) and nanoparticles ($0D$). Spin manipulation in such nanostructures is one route among many others to quantum computing. As a first step, one needs to contact a single nanoparticle and understand the interplay between Coulomb blockade and nanomagnetism, the so called magneto-Coulomb effects. Using an in-house developed near field nanolithography technique it has been made possible to contact a single nanometer sized (2-3nm) Au cluster to ferromagnetic electrodes and demonstrate a significant enhancement

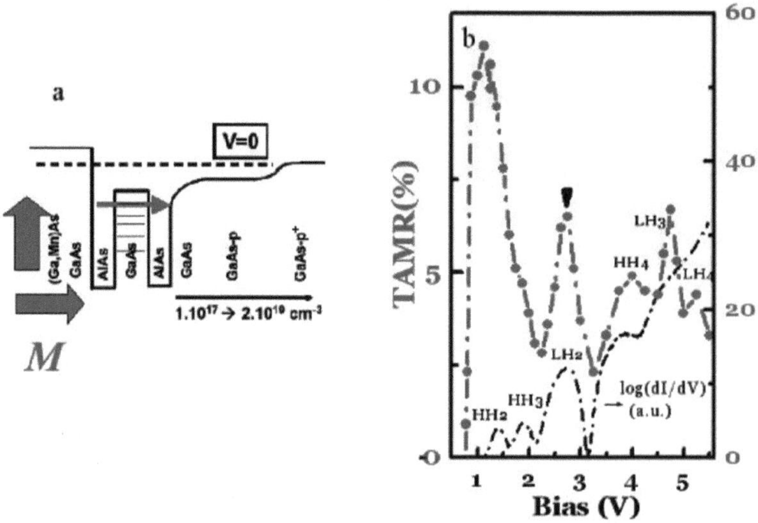

FIGURE 3. a) Schematic representation of the sample structure. b) $TAMR - V$ signal (straight red line) and $dI/dV - V$ (dashed line) acquired on the $6 - nmGaAs$ QW. The left axis displays the measured $TAMR$; the right one takes into account the up-renormalization due to the additional resistance from the depletion $GaAs$ region.

of the spin lifetime on the nanoparticle [18]. Moreover, it has been shown that the magnetic anisotropy of the ferromagnetic electrodes can affect strongly the energetic state of the particle in a way similar to a gate effect. Studies are now being extended to other nano-objects such as nanomagnets, molecules, and semi-conducting quantum dots. An other important step is related to the spin transfer

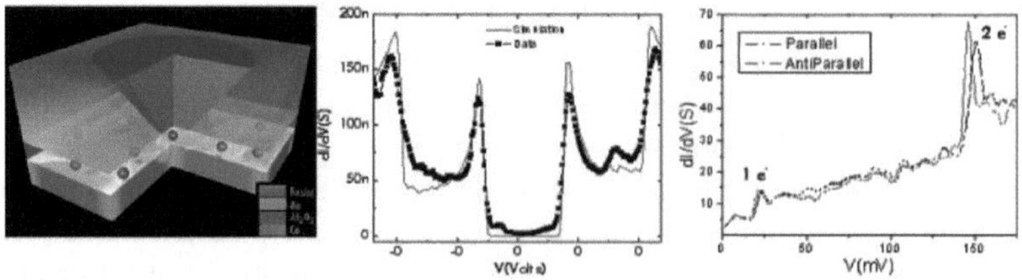

FIGURE 4. (left) Schematic view of the studied $Co/Al_2 0_3/AuNanopart./Al_2 0_3/Co$ structure. (center) I(V) curve showing Coulomb-blockade. (right) I(V) curve shifting due to spin accumulation.

phenomenon. In a spin transfer phenomenon, one manipulates the magnetic moment of a ferromagnetic body without applying any magnetic field but only by a transfer of spin angular momentum from a spin-polarized electrical current. The result of this transfer of spin can be described as a torque acting on the magnetization (Spin Transfer Torque or STT). This torque can be used either to rotate and switch the magnetization, or to displace a domain wall separating two magnetic domains, or to generate oscillations in the microwave frequency range. Originally predicted by John Slonczewski from IBM, a group in Cornell in 2000 and the group of Fert in 2001 [19] were the first to observe the reversal of a magnetization by spin transfer-induced precessions. These results triggered an intense activity of research and, today, the precessional magnetic switching by spin transfer torque is mastered in several types of magnetic devices, metallic multilayers, spin valves or tunnel junctions. As discussed in the next section, magnetic switching by STT, that is by an electronically induced precession without any need of magnetic field, will replace the switching by an applied field used in today's MRAM.

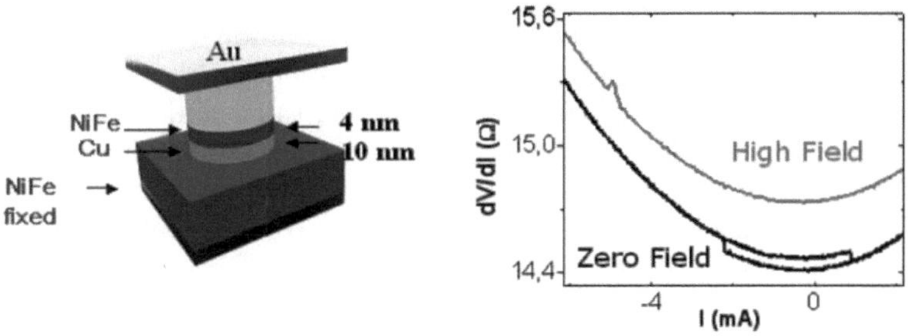

FIGURE 5. (left) Representative view of a spin valve nanopillar for spin transfer torque experiments composed of a *NiFe* layer with a fixed magnetization separated from a second thin *NiFe* layer by a nonmagnetic Cu layer. (right) Representative curves of the differential resistance vs injected current in the two regime of spin transfer phenomena : at zero applied field, hysteretic curve and at high field, reversible peak only for one current polarity.

3. Present and future technological impact of GMR and spintronics

Most applications of the GMR are based on trilayer structures of spin valve (SV) type, in which the switching between magnetic states (antiparallel and parallel magnetizations) is obtained at very low field by the rotation of the magnetization of a free magnetic layer. The concept was initially described in Grünberg's 1988 patent and, with some additional aspects, in an IBM patent of 1993 (US patent

52065590r). GMR read heads for hard discs drives (HDD) were introduced onto the market by IBM in 1997 only 8 years after the discovery of GMR. In terms of economical impact, about 5 billion GMR read-heads have been produced in total as for 2007 with a worldwide annual revenue larger than 31 billion \$. The sensitivity of the GMR to detect small fields has led to a huge increase of the density of magnetic information stored in a disc.The compound growth rate of storage areal density has increased up to 100% per year (doubling every in the period 1997-2002), and the areal density is now around $200 Gbit/in^2$, which corresponds to an increase by more two orders of magnitude with respect to the pre-GMR HDD. This opened the way to both smaller HDD forms factors (down to 0.85 disc diameter!) for mobile appliances such as ultra laptops or portable multimedia players, and to record HDD capacities (up to 1 Tbyte!). More generally, the availability of cheap HDD with high capacity ($> 100GB$) has triggered today's pervasiveness of hard disk drives in consumer's electronics such as portable MP3 players, hard-disk-video recorders, automotive HDDs, and digital cameras to list a few. HDD are also now replacing tape in at least the first tiers of data archival strategies, for which they provide faster random access and higher data rates. A transition to TMR heads began in 2004 (Seagate) with volumes now at about 300 million vs GMR heads of 500 million (2007 projected, see Figure 6). The TMR provides a higher sensitivity than the conventional GMR but sensors have a large resistance and the increase of this resistance at smaller size is a disadvantage for a further downscaling of the heads. Going forward to reach an areal density above the $Tbit/in^2$, a transition back to all metallic GMR multilayer devices is likely due to favourable sheet resistance scaling with smaller size. These heads will exploit the CPP-GMR effect.

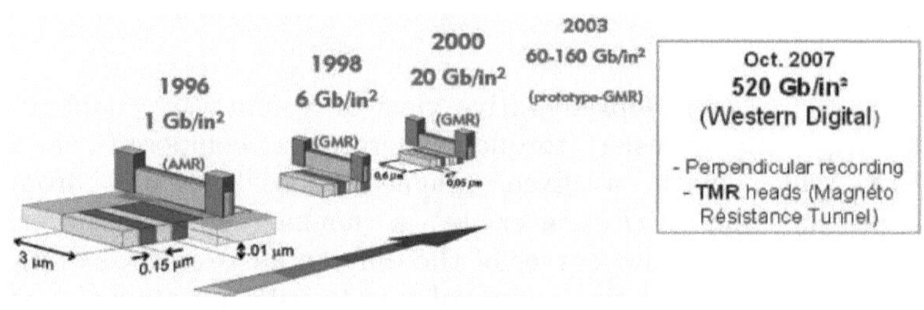

FIGURE 6. Evolution of the storage density : from Anositropic MagnetoResistance to Giant MagnetoResistance.

Actually GMR/TMR has many other applications than the HDD. A very important application relates to non volatile memories called MRAM (Magnetic Random Access Memory). The MRAMs, with their combination of the non volatility (they retain data after the power is turned off), a fast random access similar to that of the semiconductor-based RAM (5 ns read/write has been demonstrated for MRAM) and an infinite endurance (against 105 cycles for a Flash), are likely candidates for becoming the universal memory, one of the Holy Grails of nanoelectronics. The first generation of MRAM was developed with alumina-based MTJ.

The first product, a 4 Mbit standalone memory, was commercialized by Freescale in 2006 and soon voted Product of the Year by Electronics Products Magazine. Moreover Freescale introduced in June 2007 a new version able to work in the enhanced $40°C$ to $105°C$ temperature range, i.e. qualifying for military and space applications where the MRAM will also benefit from the intrinsic radiation hardness of magnetic storage. However, a major limitation to ultrahigh density in this first generation of MRAM comes from the relatively low TMR of the alumina-based MTJ and from the resulting requirement of one transistor per MTJ. Another limitation comes from the writing by magnetic fields generated by current lines. With this type of writing, the needed very large current densities and the unavoidable distribution of the writing parameters also limits the downscaling of the device. These limitations will be turned away in the next generation of MRAM using the very high TMR of MgO-based MTJ and a new concept of writing based on the STT mechanism (both described above). Several companies have already presented very promising demonstrators of this new type of MRAM which has been called STT-RAM.

GMR/TMR has also a significant impact on magnetic field sensing devices, used in a number of very competitive markets. GMR-based magnetic field sensors for the automotive industry have been on the market since 1993. GMR/TMR sensors are also used now for biomedical applications, for example in the analysis of biomolecules (ultra-sensitive detection of molecules once they have been labeled with magnetic particles, see, for example [20], and show promises for biomedical imaging such as magnetocardiography and magnetoencephalography [21].

Last but not least, spin transfer devices will certainly have also important applications in the technology of telecommunications for the generation of oscillations in the microwave frequency range. The so-called Spin Transfer Oscillators (STO) have interesting advantages, in particular the easy and fast control (agility) of the frequency by tuning a DC current (see Figure 7). Up to recently, they had the disadvantage of needing an applied field but the group of Fert has now conceived and operated a special type of STO working at zero field [22]. As, at the moment, the microwave power of the STOs is still too small for applications, synchronizing an array of STOs to increase this power is now the crucial challenge before developing practical devices. Fert's team is one of the three groups in the world having already obtained successful theoretical [23] and experimental results [24] on this problem of synchronization of STOs.

In summary, the pioneering research of Albert Fert and Peter Grünberg on spin dependent transport and magnetic multilayers led them to the discovery of the GMR in 1988. Following this discovery, the new field of spintronics has undergone outstanding progress in twenty years, under the convergence of a chain of scientific breakthroughs and technology advances. Traditional hard disk recording has gained orders of magnitude in storage capacity, thus entering the consumer electronics market. New applications have or will appear, yet only for niche markets, but expected to progress and impact rapidly. On a more long term, the use of

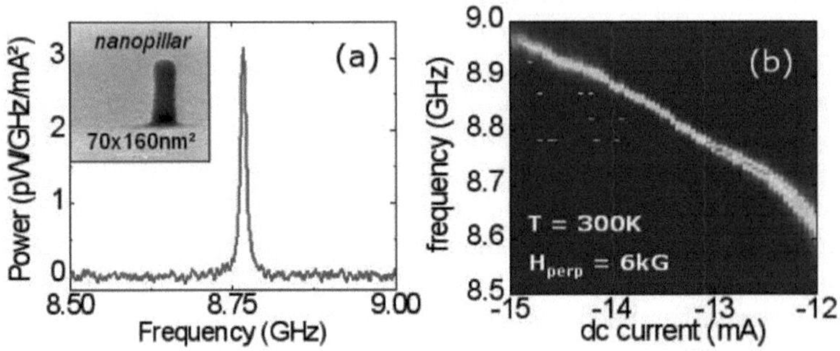

FIGURE 7. (left) High frequency peak due to the spin transfer induced emission in a spin valve nanopilar (see inset) recorded at room temperature for $I = -13mA$ and $H = 6kG$. (right) Variation of the emission frequency as a function of the injected current (the color power scale indicates the amplitude of the power).

spin injection and spin currents may lead to spin logic devices, a promising route for nanoelectronics.

Acknowledgments. The past and recent research at the Unité Mixte de Physique CNRS/Thales was done in the frame of numerous national and international collaborations, that could not be all cited here but we would like to thank all our collaborators here.

References

[1] A. Fert and I.A. Campbell, Phys. Rev. Lett. **21** (1968), 1190 ; J. Phys. **F 6** (1976), 849.

[2] P. Grünberg et al., Phys. Rev. Lett. **57** (1986), 2442.

[3] Baibich et al., Phys. Rev. Lett. **61** (1988), 2412.

[4] Binash et al., Phys. Rev. B **39** (1989), 4828.

[5] Mosca et al., J. Magn. Magn. Mater. **94** (1991), L1.

[6] S. Parkin et al., PRL **66** (1991), 2152.

[7] P. Levy and S.F. Zhang, Phys. Rev. Lett. **65** (1990), 1643.

[8] T. Valet and A. Fert, Phys. Rev. B **48** (1993), 7099.

[9] A. Fert et al., IEEE Transactions on Electron Devices **54** (5) (2007), 921.

[10] L. Hueso et al., Nature **445** (2007), 410.

[11] P. Günberg, Europhys. Lett. **66** (2004), 736.

[12] M. Bowen et al., Appl. Phys. Lett. **82** (2003), 233.

[13] M. Bowen et al., Appl. Phys. Lett. **79** (2001), 1655.

[14] Garcia et al., Phys. Rev. Lett. **97** (2006), 246802.
[15] R. Mattana et al., Phys. Rev. Lett. **90** (2003), 166601.
[16] M. Elsen et al., Phys. Rev. B **73** (2006), 035303.
[17] M. Elsen et al., Phys. Rev. Lett. **99** (2007), 127203.
[18] P. Sénéor et al., J. Phys.: Condens. Matter **19** (2007), 165222.
[19] J. Grollier et al., Appl. Phys. Lett. **78** (2001), 3663.
[20] D.K. Wood et al., Sensors and Actuators A**120** (2005), 1.
[21] R.C. Chaves et al., Appl. Phys. Lett. **91** (2007), 102504.
[22] O. Boulle et al., Nature Physics **3** (2007), 492 ; Phys. Rev. B **77** (2008), 174403.
[23] J. Grollier et al., Phys. Rev. B **73** (2006), 060409.
[24] B. Georges et al., Phys. Rev. Lett. (2008).

Vincent Cros
Albert Fert
Pierre Sénéor
Frédéric Petroff
Unité mixte de Physique CNRS/Thales
et Université Paris Sud
Palaiseau et Orsay, France

The Spin
B. Duplantier, J.-M. Raimond and V. Rivasseau, Eds., 159–193

Poincaré Seminar 2007

Magnetic Resonance Imaging: From Spin Physics to Medical Diagnosis

Pierre-Jean Nacher

Abstract. Two rather similar historical evolutions are evoked, each one originating in fundamental spin studies by physicists, and ending as magnetic resonance imaging (MRI), a set of invaluable tools for clinical diagnosis in the hands of medical doctors. The first one starts with the early work on nuclear magnetic resonance, the founding stone of the usual proton-based MRI, of which the basic principles are described. The second one starts with the optical pumping developments made to study the effects of spin polarization in various fundamental problems. Its unexpected outcome is a unique imaging modality, also based on MRI, for the study of lung physiology and pathologies.

1. Historical introduction

Magnetic Resonance Imaging (MRI), now widely known for its usefulness as a medical diagnosis tool and for the variety of clear pictures of the body's interior obtained in a harmless and non-invasive manner, had its foundations laid more than 60 years ago in physics experiments designed to measure properties of the nuclear spins of hydrogen atoms. In even earlier experiments, Rabi had shown that an oscillating magnetic field could induce transitions between levels associated to the spin state of various nuclei in an applied static magnetic field [1]. This pioneering work was performed on molecular beams, using a selection and detection method of the nuclear spin state similar to that developed in the 1920's by Stern and Gerlach for their demonstration of spin quantification in silver atoms. The transposition of Rabi's observation of Nuclear Magnetic Resonance (NMR) in a beam of independent molecules to solid and liquid samples was successfully done independently in 1945 by Purcell and by Bloch.[1] The key feature of both experiments was the observation of the resonance phenomenon through electromagnetic

[1] An earlier attempt in Leiden to observe NMR in solids had failed because of exceptionally long relaxation times [9]. Gorter first used the term "nuclear magnetic resonance" in this publication,

detection at the resonance frequency. At the M.I.T.,[2] Purcell, Torrey, and Pound, worked on solid paraffin filling a 1-liter cavity tuned at 30 MHz. They observed a 0.4% change in the rf signal amplitude fed to the cavity due to energy dissipation arising from nuclear spin relaxation of H atoms [2]. Meanwhile, at Stanford University, Bloch, Hansen, and Packard, performed similar experiments at 7.7 MHz on a 1.5-cc sample of water. They used two orthogonal rf coils: the receive coil detected the re-emission of the resonant rf power absorbed from the transmit coil by the nuclei of the water protons (H atoms) [3, 4]. Although Rabi's work was crucial as the initial demonstration of NMR (he was awarded the Physics Nobel prize in 1944 "for his resonance method for recording the magnetic properties of atomic nuclei"), the conceptual and technical leap achieved by Bloch and Purcell really opened the way to modern NMR and MRI. They were awarded the 1952 Physics Nobel prize "for their development of new methods for nuclear magnetic precision measurements and discoveries in connection therewith". A last essential tool of NMR was found independently by Torrey [5] and Hahn [6], who demonstrated the feasibility of pulsed NMR (initially suggested by Bloch) and observed free Larmor precession. Hahn further used pulsed NMR to generate and observe spin echoes [7].

The next 20 years saw the development of NMR as a powerful investigative tool in many areas of physics and even more so in chemistry. The sensitivity of the nucleus to its electronic environment in a molecule (the "chemical shift") and spin-spin interactions were seen at first in the nuclear physics community as annoying features, but the potential of NMR spectroscopy for analytical studies was soon revealed by the discovery of the 3 peaks of ethanol in Purcell's group [8]. Almost none of the early applications of NMR were medical, although a large amount of work was published on relaxation, diffusion, and exchange of water in cells and tissues, even in living human subjects [10] and whole animals [11].

However it was not until 1973 that NMR was used to generate true (2-D) images, when Lauterbur, after his early work on chemical shifts, produced images (of two glass tubes filled with water) reconstructed from a series of 1-D projections obtained using a magnetic field gradient [12]. Soon, MRI benefited from essential developments performed by Mansfield in Nottingham to efficiently generate images, such as slice selection [13] and fast acquisition schemes [14]. These pioneering contributions were jointly rewarded by the 2003 Nobel prize in Medicine "for their discoveries concerning magnetic resonance imaging". Another essential contribution to MRI originated from Zurich, where Ernst applied his Fourier spectroscopy techniques [15] to imaging [16] (Ernst was awarded the Nobel Prize in

attributing the coining of the expression to Rabi. In Kazan, another place where major contributions to NMR were made, Zavoiskii had also failed to reliably detect NMR in 1941, but he discovered electron spin resonance (EPR) in 1944.

[2]Purcell obtained his PhD degree in 1938 at Havard University, where he returned in 1945, became Professor of Physics in 1949, and performed most of his work. But he spent the war years in the Fundamental Developments Group in the Radiation Laboratory of the M.I.T., where he was associated in particular to Rabi. Purcell's seminal publication on NMR [2] bears the M.I.T. address.

Chemistry in 1991 "for his contributions to the development of the methodology of high resolution nuclear magnetic resonance spectroscopy").

Research in MRI was pursued only in a few academic laboratories, mostly in the UK, with the first image of a human thorax in 1977, of a head in 1978, until 1980 when Edelstein, in Mallard's team in Aberdeen, obtained the first clinically useful image of a patient [17]. After this date, several companies started to invest in MRI developments and to promote clinical trials. For instance, Toshiba marketed the first commercial MRI scanner in 1983 (with a 0.15 T field), while General Electric, the current leading manufacturer, enrolled several of the MRI pioneers, including Edelstein, and produced the first 1.5 T clinical system in 1985. With the growing availability and performance of MRI instruments, scientific publications on MRI in medical journals linearly increased in number from 1984, to reach 16000 per year in 2006 [18]. At the same time, MRI exams progressively became a routine diagnosis tool, with now 70 million exams per year and 40 thousand machines in the world.

The rest of this paper is organized as follows. The next section briefly provides the simple notions required to understand the usual NMR dynamics involved in MRI applications. The following section describes the basic principles of imaging, and the influence of elementary physical processes such as relaxation and molecular diffusion on image quality. Finally, the last section is devoted to an unexpected application of decades of developments in the field of polarized noble gas production. This work, that originally aimed at providing tools for various fundamental studies involving spin physics, now also provides an unrivaled tool for MRI of the lung.

2. Basic physics of NMR

2.1. Bloch's equations and NMR dynamics

The phenomenon of magnetic resonance results from the dynamics of spins in combined static and oscillating magnetic fields. Since, by far, most NMR and MRI applications involve spin 1/2 nuclei, only such spin systems will be considered in the following. NMR indeed also operates for nuclei with higher spin values, but has more complex features. The quantum-mechanical derivation of spin dynamics can be found in many textbooks, and the main results are only recalled here.

In an applied static magnetic field B_0, the two eigenstates that correspond to the projections $\pm\hbar/2$ of the angular momentum on the quantization axis defined by B_0 have an energy difference:

$$\Delta E = E_- - E_+ = \hbar\omega_0 = \hbar\gamma B_0, \tag{1}$$

where the transition (Larmor) angular frequency ω_0 usually falls in the radio-frequency (rf) domain, and γ is the gyromagnetic ratio associated with the spin of the considered nucleus. For protons, the nuclei of H atoms, $\gamma/2\pi$=42.58 MHz/T. At thermal equilibrium (for a temperature T), the system is described by a diagonal density matrix with a ratio of populations:

$$N_-/N_+ = \exp(-\hbar\gamma B_0/k_{\mathrm{B}}T). \tag{2}$$

For a positive gyromagnetic ratio (such as that of protons or ^{13}C for instance, but contrary to the case of the spin 1/2 noble gases ^{3}He and ^{129}Xe that will be considered in section 4), the more populated low-energy state is that with the nuclear spin aligned in the direction of the applied field.

Transitions can be driven between these two eigenstates by a transverse rf field B_1(i.e., perpendicular to B_0) rotating at angular frequency ω, and the probability that the system is in the initial state at time t oscillates according to the Rabi formula:

$$\mathcal{P}(t) = 1 - \frac{\omega_1^2}{\omega_1^2 + (\omega - \omega_0)^2} \sin^2\left(\frac{t}{2}\sqrt{\omega_1^2 + (\omega - \omega_0)^2}\right) \tag{3}$$

where the amplitude of the rf field, $\omega_1 = \gamma B_1$ (in angular frequency units), and the detuning from resonance, $\omega - \omega_0$, determine the characteristic features of the oscillation (see Fig. 1, left)[3]. In the historical Rabi experiment [1], where the fraction of atoms remaining in the initially prepared state is measured after the action of the rf field for a variable time depending on the molecular velocity, the observed resonance curve is indeed described by Eq. 3, simply replacing the oscillating $\sin^2$ term by its time-averaged value, 1/2. An additional effect of the applied resonant rf field is to periodically create coherences (off-diagonal terms) in the density matrix describing the spin 1/2 system. They correspond to transverse components of the angular momentum, which have a time evolution best described in the rotating frame synchronous with the applied rf field (see Fig. 1, right).[4] An equivalent classical description of the spin dynamics is obtained if one considers a system with an angular momentum $\mathbf{j}$ and a magnetic moment $\mathbf{m}$ such that $\mathbf{m} = \gamma\mathbf{j}$. This vector proportionality[5] is responsible for the gyroscopic-like response of spins to an applied magnetic field, similar to the dynamics of a spinning top in a gravity field. In a physical sample, the time evolution of the macroscopic local magnetization density $\mathbf{M}$ that results from the magnetic moments of the nuclei behaves in the same way, and thus obeys Bloch's equation [4]. In fact, in most cases, this classical description can be extended to interacting spin systems by simply introducing phenomenological relaxation terms to take into account the stochastic

[3]If the transverse field is oscillating, it can be considered as composed of two counter-rotating components, among which only one (with half of the total amplitude) is resonant, whereas the other has no effect.

[4]Since any two-level quantum system is formally equivalent to a spin 1/2 system, such representation is used in many physical problems, even if the levels correspond to electronic states of an atom, connected by an electric dipolar transition. Rydberg atoms in a box are near-ideal examples of such two-level systems [19]. NMR concepts and vocabulary, such as $\pi/2$ or π pulses, are used, and the (optical) Bloch equations rules the evolution of the system.

[5]This non-classical relation does not hold, for instance, for the magnetized needle of a compass. The magnetic moment is a constant locked to the long axis of the needle, but the angular momentum is proportional to the angular frequency of rotation of the needle.

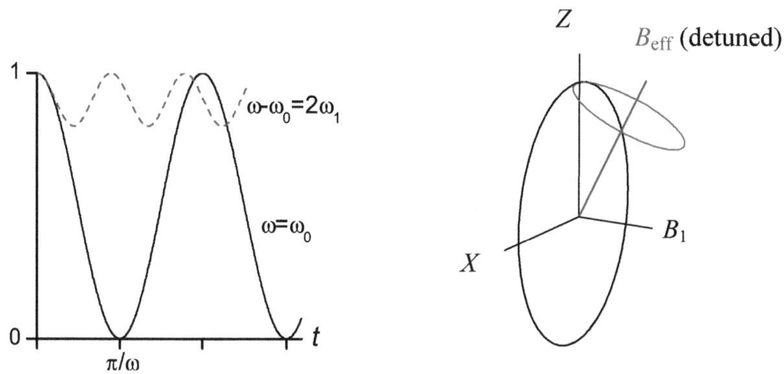

FIGURE 1. **Left:** time evolution of the probability for a spin 1/2 to be in the initial eigenstate under rf irradiation (Eq. 3). Solid line: resonant rf, $\omega{=}\omega_0$; dashed line: detuned rf , $\omega - \omega_0 = 2\omega_1$; ω_0 is the resonance (Larmor) angular frequency, ω_1 is the amplitude of the rotating rf component in angular frequency units. **Right:** The corresponding trajectories of the classical momentum associated with this spin are circles on the so-called Bloch unity sphere. The longitudinal (Z) component is the Rabi oscillation probability plotted on the left, and the projection in the transverse (XY) plane or represents the off-diagonal elements in the spin density matrix (Y is the direction of the rf field).

effect of dipolar spin interactions.[6] This leads to the usual Bloch's equation in its complete form:

$$\frac{d\mathbf{M}}{dt} = \gamma \mathbf{M} \times \mathbf{B}_{\text{eff}} - [R]\,\mathbf{M} - D\mathbf{\Delta}\mathbf{M}. \tag{4}$$

It is most conveniently written in the rotating frame synchronous with the rf field. The effective field in the first term of Eq. 4 is then given by:

$$\mathbf{B}_{\text{eff}} = -\omega_1 \hat{x}/\gamma - (\omega - \omega_0)\,\hat{z}/\gamma. \tag{5}$$

When no rf field is applied, a frame rotating at the local Larmor frequency is most convenient (then $\mathbf{B}_{\text{eff}}{=}0$), but a frame rotating at any frequency ω, e.g. the average Larmor frequency over the extended physical sample, can also advantageously be used (then $\mathbf{B}_{\text{eff}}$ is given by Eq. 5 with $\omega_1{=}0$). The second term in Eq. 4 is a relaxation term characterized by two different rates: one for the longitudinal component M_z of $\mathbf{M}$, noted $1/T_1$, and one for the transverse components, noted

[6] A full quantum treatment of the spin dynamics is required only in rare particular cases, for instance when short-range quantum correlations between interacting spins of nuclei in a molecule play a key role. Such unusual situations will not be considered here, and the classical description will always be used. A discussion regarding the necessity of using a quantum formalism can be found in [23] and references therein.

$1/T_2$. The last term in Eq. 4 describes the irreversible effect of atomic diffusion, that depends on the spin diffusion coefficient D.

In cw NMR with weak rf fields, relaxation phenomena play an important role to determine the response of the spin system to irradiation, and hence deeply influence the observed spectra. In contrast, in pulsed NMR, as performed for MR imaging (see section 3), the rf field is usually so intense that relaxation and diffusion phenomena have no significant influence during the pulse. Hence the right hand side of Bloch's equation (Eq. 4) is reduced to its first term, and $\mathbf{M}$ simply undergoes a rotation around the direction of the effective field. At resonance, any tip angle with respect to the initially longitudinal orientation of $\mathbf{M}$ can be achieved using an appropriately timed tipping rf pulse (see Fig. 1, right). Finally, in the absence of applied rf (e.g., during the free evolution following a tipping pulse), the complete Bloch's equation shows that a uniform magnetization in a uniform field B_0 simply precesses at Larmor frequency around the field axis, with a transverse component that decays exponentially (with a rate $1/T_2$) and a longitudinal component that exponentially recovers its equilibrium value, determined by Eq. 2 (with a rate $1/T_1$). If the magnetic field B_0 is not uniform (as is inevitably the case in an experiment), the transverse part $\mathbf{M}_\perp$ of $\mathbf{M}$ progressively acquires a non-uniform phase, and magnetization currents induce an additional decay of $\mathbf{M}_\perp$ according to the third term in Bloch's equation.

2.2. Signal amplitude considerations

In the historical Rabi experiment, the resonance was observed by monitoring the flux of molecules in a weak but fully polarized beam. Similarly, optical detection of magnetic resonance in optically pumped atomic vapors, proposed by Kastler[7] [21] and demonstrated by Brossel and Cagnac [22], was a very sensitive method to observe magnetic resonance phenomena in very dilute systems. The common feature of these approaches is that the system's polarization is complete or very high, and that the detection of atoms or of visible photons, with energies in the eV range, can provide excellent signal-to-noise ratios (SNRs) even with a modest number of detected events. In contrast, standard NMR with radioelectric detection of the resonance, as pioneered by Purcell and Bloch in their seminal work, relies on the detection of (magnetic) rf oscillations involving much lower energies (NMR frequencies are in the MHz range, 10^{-9} times lower than optical frequencies). Moreover it is performed in weakly polarized samples (from Eq. 2, equilibrium polarization is of order 10^{-5} at room temperature for B_0=1T). Altogether, a dramatic loss of sensitivity (of 14 orders of magnitude) is expected to result from these combined reductions in signal amplitude. Fortunately, this is compensated by a huge increase in density from molecular beams or dilute atomic vapors to liquid or solid samples.

[7] A. Kastler was awarded the 1966 Nobel prize in physics "for the discovery and development of optical methods for studying Hertzian resonances in atoms". The principle of optical pumping, and its application to the obtention of nuclear polarization in noble gases, will be described in section 4.

Indeed, such a discussion on signal amplitudes only is rather crude, and more careful SNR arguments must be considered. NMR signal during free precession is usually picked up by a coil that converts the oscillating flux of precessing magnetization into a recorded electromotive force (emf). For given sample and coil geometries (their effect is not discussed here), the emf simply scales with B_0^2 because the equilibrium magnetization is proportional to B_0, and because Faraday's law introduces a time derivative, hence a factor ω_0. Estimating the field dependence of noise is more delicate, and implies making assumptions on its physical origin. Disregarding instrumental noise arising from rf interference (efficiently suppressed when operating in a Faraday cage or shielded room) or generated by amplifiers and recording electronics, unavoidable noise originates from the coil and from the sample. This thermal (Johnson-Nyquist) noise results from thermal agitation of charge carriers, in the coil and in the sample. For the latter, Faraday's law introduces the same factor ω_0 into the induced emf as in does for the signal, so whenever noise mostly originates from the sample, SNR increases only linearly with B_0. In contrast, whenever noise mostly originates from the coil, it can be argued that SNR increases more rapidly with the operating field, scaling as $B_0^{7/4}$ [24]. For a well-designed coil and for medical MRI applications, the sample (i.e., the patient's tissues) is the dominating source of noise for sample sizes exceeding a few centimeters. Conversely, coil noise dominates for non-conducting samples, for MRI of small animals or for MR microscopy using small-sized surface coils. In this case, it can be efficiently reduced using cold probes, or even superconducting coils [25]. The thermal noise spectral density, that is given by $\sqrt{4k_BTR}$ (across a resistor R at temperature T), is thus decreased through the reduction both of R and T. More exotic options are being explored to further improve SNR in MRI, such as the use of SQUID-based devices [26] or of optical magnetometers [27] at very low field, or of force detection in magnetic resonance force microscopy for sub-micron resolution MRI [28].

3. Principles of MRI

3.1. Effect of an applied field gradient: 1-D imaging

When NMR is performed in a position-dependent magnetic field $B_0(\mathbf{r})$, the Larmor precession frequency correspondingly depends on position $\mathbf{r}$ in the sample. An important practical case is met when a uniform field gradient $G\hat{\mathbf{k}}$ is applied:

$$B_0(\mathbf{r}) = B_0(0) + G\hat{\mathbf{k}} \cdot \mathbf{r}. \tag{6}$$

The precession frequency only depends on position along the direction $\hat{\mathbf{k}}$ of the gradient, and resonance synchronously occurs in any plane perpendicular to $\hat{\mathbf{k}}$. If cw NMR is performed, signal at a given frequency is induced by atoms the vicinity of one such plane (there is in fact a limit to the accuracy of the localization resulting from the NMR intrinsic line width for G=0). If the NMR signal is received using a coil with uniform coupling to all parts of the sample (e.g., using a long solenoidal coil), NMR spectra recorded when sweeping the field or the frequency represent 1-D projections of the magnetization density in the sample. The frequency scale is set by the amplitude G of the gradient, and the projection is actually convolved with the Lorentzian line shape corresponding to the relaxation rate $1/T_2$, which limits the achievable resolution along the direction of the gradient to a scale $\delta r{=}1/\gamma G T_2$. This is illustrated in Fig. 2 for the case of a spherical sample with uniform magnetization density. In the remainder of this section, we shall assume that large values of G are used ($\gamma G\rho T_2 \gg 1$), and moreover that diffusion effects can be neglected.

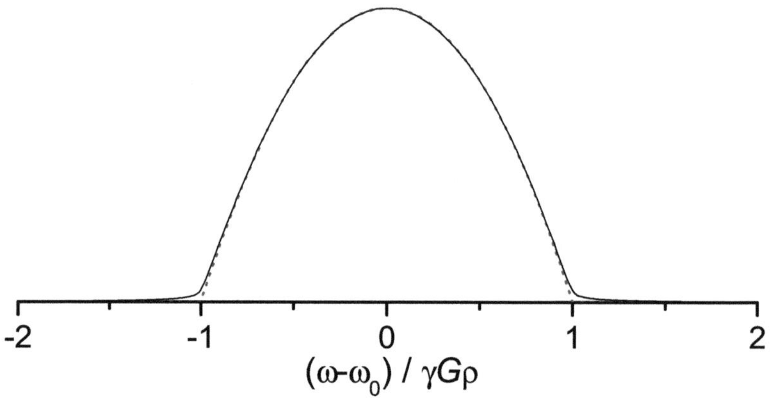

FIGURE 2. Computed cw NMR spectra for a uniform spherical sample of radius ρ in a uniform gradient of amplitude G. Without relaxation (dashed line), the spectrum is the positive part of the parabola 1-$(\omega-\omega_0)^2/(\gamma G\rho)^2$, directly mapping the sections of the sphere. With some amount of relaxation in Bloch's equation (but neglecting diffusion effects), sharp edges are smoothed out (solid line: $\gamma G\rho T_2{=}0.01$).

Pulsed NMR is actually preferred to the very time-consuming cw NMR for imaging applications. Following a tipping pulse, the free-precessing magnetization induces an emf in the coil:

$$e(t) \propto \int_{sample} M(\hat{\mathbf{k}}\cdot\mathbf{r}) \cos\left[\gamma B_0(\mathbf{r})t + \varphi\right] e^{-t/T_2}, \tag{7}$$

where the local value of the field (Eq. 7) sets the local Larmor frequency, and φ is a phase term depending on the tipping pulse and on the coil position. The actually recorded NMR signal is obtained by beating the high-frequency voltage $e(t)$ with a reference voltage at a fixed frequency ω_{ref}, e.g. $\gamma B_0(0)$. The resulting in-phase and quadrature voltages are the components of a complex signal $S(t)$:

$$S(t) = \int_{sample} M(\hat{\mathbf{k}}\cdot\mathbf{r}) \exp\left[i\gamma t G\hat{\mathbf{k}}\cdot\mathbf{r}\right] e^{-t/T_2} \tag{8}$$

that represents the sum of contributions from all parts of the sample in the frame rotating at the reference frequency ω_{ref}. This complex signal keeps track of the sense of rotation in the rotating frame, and thus discriminates between frequencies higher and lower than ω_{ref}. The time evolution of S is illustrated in Fig. 3 for the same spherical sample as in Fig. 2. A Fourier transform (FT) allows one to switch

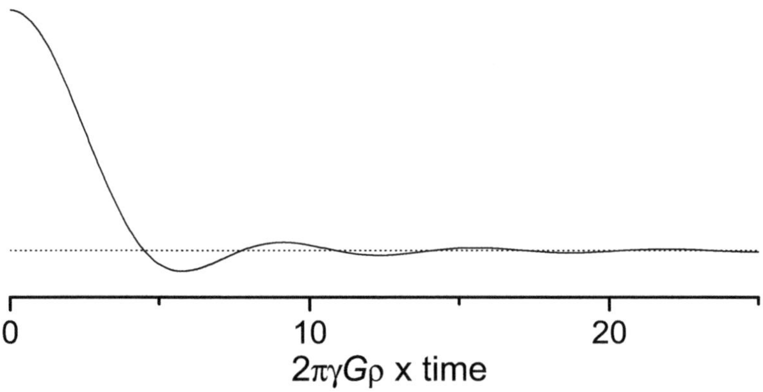

FIGURE 3. Computed NMR signal (Eq. 8) for a uniform spherical sample of radius ρ in a uniform gradient of amplitude G (neglecting relaxation and diffusion effects). The signal has a constant phase, and the quadrature (dotted line) remains null at all times.

from the time domain to the frequency domain, and to obtain from this signal the spectrum that would be recorded in a cw experiment. Although the knowledge of S only at positive times is in principle sufficient to derive the spectrum, hence the 1-D projection image of the sample (computing the FT of the function defined from the signal by setting $S(-t)=S^*(t)$), gradient-echo or spin-echo techniques are often used to acquire more information in the time domain.

The fast decay of the signal, with a time scale of order $1/\gamma G\rho$ for a sample of size ρ, essentially results from the phase decoherence of the local magnetization. Since the local magnetization has hardly decayed under the action of relaxation, the phases can be refocused, and a so-called gradient echo can be formed with a large value of the signal recovered at a finite time after the tipping pulse. Phase refocusing can result from an inversion of the gradient at time τ, after which the signal is given by:

$$S(t > \tau) = \int_{sample} M(\hat{\mathbf{k}} \cdot \mathbf{r}) \exp\left[i\gamma\,(2\tau - t)\,G\hat{\mathbf{k}} \cdot \mathbf{r}\right] e^{-t/T_2}. \tag{9}$$

The initially fast precessing spins become slow precessing ones after the gradient sign change, and all phases are back to their initial values at time 2τ, i.e. the magnetization is uniform again. Following this, evolution proceeds as it does just following a tipping pulse (see Fig. 3). A more general time variation of G can be used, with different amplitudes before and after the time τ. G can also be periodically reversed at times τ, 3τ, 5τ,...: an echo is formed whenever the time integral of $G(t)$ vanishes and an echo train is obtained.

Another method to obtain an echo consists in leaving the gradient unchanged, but in inverting the sign of all phases at time τ using a π rotation around any direction in the transverse plane. The phase advance of the "fast" spins suddenly becomes a phase lag, and rephasing occurs again at time 2τ. These so-called spin echoes can also be periodically refocused by repeating the rf pulse inducing the π rotation.

Signals obtained using a spin echo or a gradient echo technique are similar, but experimental constraints or imperfections may support the choice of either technique. The main drawback of the gradient echo technique is that gradients can only be approximately reversed. Indeed, the applied gradient that results from current fed to a dedicated set of coils (the gradient coils) can be reversed. But the unavoidable static field map imperfections (that are usually not a uniform gradient) induce a progressive loss of phase coherence that is unaffected by the inversion of G at time τ. Altogether, a gradient echo can be observed only during the experimental coherence time of the signal, T_2^*, that is often shorter than the transverse relaxation time T_2. This limitation is removed by the use of a spin echo technique, since the dephasing due to all field variations (both G and the imperfections of B_0) is exactly refocused. However frequently firing intense rf pulses to induce π rotations requires a fair amount of power, which may in some cases pose safety problems due to tissue heating.

The time variation of the signal is sampled, with a sampling frequency f_s, during a time window of duration T_{obs} that is usually centered[8] on the echo time 2τ. The spectrum obtained by FT from these data has a frequency resolution $1/T_{obs}$ and extends over a frequency range $\pm f_s/2$. For the 1-D projection of the sample, the frequency resolution corresponds to a spatial resolution of $1/\gamma G T_{obs}$. Of course this is only a maximal resolution, that is reduced due to transverse relaxation time (T_2, see Fig. 2, or T_2^*), and also due to SNR limitations. The frequency range implies a so-called field of view (FOV) of $f_s/\gamma G$. The FOV must be larger than the sample size to avoid folding artifacts resulting from the undersampling of high frequencies.

3.2. 2-D and 3-D imaging methods

Various strategies can be used to obtain information on distribution of magnetization in a sample, based on extensions of the 1-D projection imaging method.

[8]This is not in fact mandatory. Using symmetry properties of echoes that result from the fact that the magnetization phase is uniform at t=0, half of the echo is sufficient to provide the information needed to retrieve spectral data. It is only important to include the echo time 2τ in the recording. It is technically difficult in simple pulsed NMR to have the full decay from t=0 due to the delay for recovery of saturated detection electronics following the rf tipping pulse. A symmetric echo is usually recorded to increase SNR, but asymmetric echoes are used is ultra-fast acquisition schemes.

A widely used technique is the selective rf excitation of only part of the sample, or slice selection. If a rf tipping pulse is applied in the presence of a gradient $G\hat{\mathbf{k}}$, it is non-resonant, and has thus no tipping action, over all of the sample but in the plane defined by $\hat{\mathbf{k}} \cdot \mathbf{r} = (\omega - \omega_0)/\gamma G$. The time duration of the rf pulse determines the thickness of the slice in which significant tipping occurs, and the time envelope of its rf amplitude sets the variation of the tip angle with position. For small tip angles, these are indeed linked by FT relations, and uniform tipping in a slice of given thickness is obtained using a sinc-shaped rf pulse. More complicated pulse shaping [29] is required at large tip angle because of the non-linear variation of the angle with rf amplitude and detuning (see Fig. 1). 3-D imaging of the sample can be performed combining slice selection, and 2-D imaging in each selectively excited plane.

The standard 2-D cartesian imaging technique consists in acquiring a series of gradient echoes obtained in an applied (so-called readout) gradient $\mathbf{G}_\mathrm{r}$, using a variable (so-called phase-encoding) gradient $\mathbf{G}_\mathrm{e}$ applied along a perpendicular direction before the echo is obtained. For simplicity, we assume here that $\mathbf{G}_\mathrm{r}$ is oriented along the x-axis, and $\mathbf{G}_\mathrm{e}$ along the y-axis, but all orientations can be freely chosen. The recorded signal is thus modified with respect to Eq. 9 and reads:

$$S(t > \tau) = \int_{sample} M(x, y) \exp\left[i\gamma\tau_\mathrm{e} G_\mathrm{e} y\right] \exp\left[i\gamma\left(2\tau - t\right) G_\mathrm{r} x\right] e^{-t/T_2}. \qquad (10)$$

For each value of the phase-encoding parameter $\tau_\mathrm{e} G_\mathrm{e}$, a Fourier component of the spatial modulation of the magnetization in the sample along the y-axis is probed (see Fig. 4, left). With N_e data acquisitions performed for evenly-spaced gradients amplitudes between -G_max and G_max, the so-called k-space (here a 2-D array of data filled recorded time-varying signals) contains sufficient information to compute a 2-D image by (discrete) inverse FT of the data. The FOV in the physical direction of the readout gradient is given by the formula of the previous section 3.1, while in the direction of the encoding gradient the FOV given by $N_\mathrm{e}/2\gamma\tau_\mathrm{e} G_\mathrm{max}$.

The original radial projection method introduced by Lauterbur [12] can be used with a series of gradient-echo acquisition of data, for successive gradient orientations $\hat{\mathbf{k}}$ sampling a plane (see Fig. 4). Image reconstruction can be performed using back-projection algorithms, or by considering that the acquired data provide enough information on the 2-D FT of the image (the k-space) to perform an inverse FT after regridding and interpolation. Radial acquisition is often preferred when fast movements are recorded: at the time of each echo, the center of the Fourier space is re-sampled, which eliminates movement artifacts that deeply affect other imaging methods, and allows high time resolution using sliding window methods.

There are indeed other strategies to acquire k-space data, e.g. following different trajectories (spiral or interleaved spiral trajectories), or performing only partial (e.g. half-plane) acquisition [30]. Current advances in fast MRI involve both hardware and software developments. Parallel data acquisition using coil arrays to directly provide spatial information allows one to undersample the k-space [31, 32].

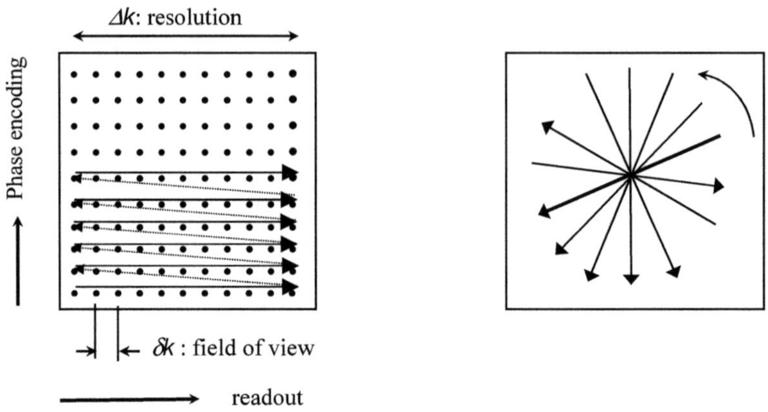

FIGURE 4. Examples of trajectories for 2-D k-space sampling. **Left:** cartesian mapping of the plane is obtained by a series of data acquisitions with steps of phase encoding in the Y direction. **Right:** radial mapping is obtained by projections onto different successive directions.

Sparse sampling of the k-space exploiting the spatial or temporal redundancy that usually allows image compression can also provide high-quality images [33]. All these techniques aim at accelerating data acquisition in order to reduce movement artifacts or to capture relevant anatomical motion (e.g. in cardiac imaging). Another motivation is that imaging sessions can take up to several tens of minutes for each patient, because of the large number of data required for high-resolution multi-slice or 3-D images, and due to the necessary signal averaging to obtain high SNR. Reducing the examination time can reduce patient discomfort and allows more cost-efficient use of the clinical MRI systems.

3.3. Contrast mechanisms

For all the above imaging strategies, the collected signal and the resulting image reflect the local magnetization density in the sample. This scales with the density of H atoms, which is high both in water and lipids of all tissues. This provides anatomical images that are usually poorly contrasted and of little clinical use. Fortunately, various physical effects can be put to use to provide enhanced contrast between different organs with similar proton density, or reveal differences between normal and pathological regions.

T_1-weighting is obtained in imaging sequences where the repetition period T_R of NMR pulses is of the order of the longitudinal relaxation time. In steady-state, the local magnetization is reduced from its thermal equilibrium value M_{eq} according to:

$$M = M_{eq} \frac{1 - \exp\left(-T_R/T_1\right)}{1 - \cos\alpha \exp\left(-T_R/T_1\right)}, \tag{11}$$

where α is the tip angle of the rf pulse. For an appropriate choice of α and T_R, parts of the sample with short T_1 will provide a stronger signal and will thus appear more intense in images than parts with a long T_1, where magnetization does not recover much between tipping pulses. In normal tissues, fat (lipids) has a shorter T_1 than water and thus appears white in T_1-weighted images (e.g., cerebral white matter appears so since it contains more lipids than grey matter). Paramagnetic contrast agents can be injected to locally induce a reduction of T_1 and thus increase signal intensity wherever blood perfusion is present.

T_2-weighting is obtained in imaging sequences where the signal acquisition time is delayed long enough after the tipping pulse (until the echo time T_e) for attenuation of the transverse magnetization (T_2 in Eq. 9 or 10) to be significant. T_2-weighted images appear reversed compared to T_1-weighted images, with bright water-containing regions and weaker signal from lipids. Both T_1- and T_2-weighted images are acquired for most medical examinations. Together, they provide the trained radiologist with a lot of morphological and functional information, since for instance flowing blood, haematomas, and various types of tumors differently affect signal intensity in MR images.

Diffusion is the last physical effect (neglected so far) that affects spin dynamics. Diffusion-weighted images are obtained by adding a bipolar gradient pulse following the tipping rf pulse before the rest of the usual imaging sequence. This first imprints a helix-like phase pattern along the direction of this added diffusion-sensitizing gradient, then unwinds it. The net result is that the transverse amplitude is reduced by a factor $\exp(-D\gamma^2G^2\tau_d^3)$, where D is the diffusion coefficient, G the amplitude and τ_d a time scale associated with the sensitizing gradient. In the human brain, water diffusion is impeded by natural barriers (cell membranes, myelin sheaths, ...) and a reduced apparent diffusion coefficient (ADC) is measured instead of D. ADC imaging of the brain is routinely performed in case of stroke (ischemic or hemmoragic). ADC mapping can also be performed as a function of the direction of the sensitizing gradient. The resulting diffusion-tensor images allow for the determination of directionality as well as the magnitude of water diffusion. This kind of MR imaging enables to visualize white matter fibers in the brain and can for instance map subtle changes in the white matter associated with diseases such as multiple sclerosis or epilepsy.

4. Lung MRI with polarized noble gases

In spite of its high abundance in tissues, H is not the only element that can be detected and imaged by NMR. Non-proton MRI, although it is not part of routine clinical exams, allows for instance useful investigation of the human heart when tuned to the frequencies of ^{31}P or ^{23}Na, which are both naturally abundant isotopes with non-zero spin. The low abundance of ^{13}C (1%) prevented carbon-based MRI until ex-vivo pre-polarization of its nuclei in selected bio-compatible molecules was used to provide dramatic signal enhancement [34, 35].

In addition to these nuclei, that are naturally present in biological tissues, the spin 1/2 noble gases ^{3}He and ^{129}Xe can be used to image the lung airways. This was first demonstrated in 1994 by a Princeton-Stony Brook collaboration using polarized xenon in excised mouse lungs [36], and was soon followed by similar demonstrations in human volunteers using polarized helium, both in the USA and in Germany [37, 38]. The striking difference between proton and ^{3}He chest images is illustrated in Fig. 5. Both images have been obtained in a 1.5 Tesla clinical

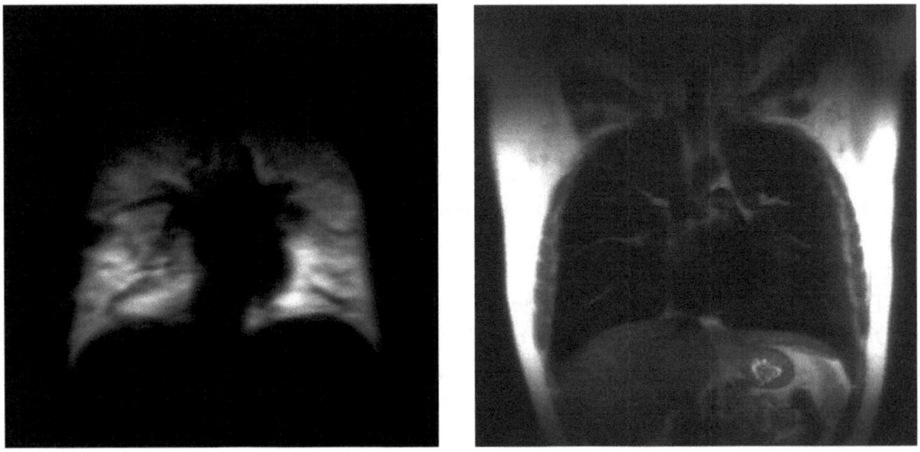

FIGURE 5. Chest MR images (1 cm thick slices) of the same normal subject acquired during breath hold. **Left:** FLASH ^{3}He image of the inhaled gas. **Right:** fast spin-echo proton image. The NMR frequency is changed from 49 MHz for ^{3}He to 64 MHz for protons. *Courtesy of Jim Wild, Academic Radiology Dept., U. of Sheffield, UK.*

MRI scanner, but using different coil systems operating at different frequencies (the gyromagnetic ratio of ^{3}He, $\gamma_{\mathrm{He}}/2\pi$=-32.44 MHz/T, is 30% lower than that of H). The left part of the figure displays a ^{3}He image recorded after the subject has inhaled a volume of order 0.3 liter of polarized gas in addition to a normal air intake. An intense signal is obtained from the lung airspaces in this slice, and a set of 15-20 such slices, recorded during a 10-12 s breath hold, allows to reconstruct a 3-D map of the gas distribution in the lungs. In this image, the space occupied by the heart is clearly visible, as well as a peripheral wedge-shape ventilation defect in the right lung, of a kind more often observed in asthmatic subjects. More details on the gas polarization process and on the specific imaging protocol will be given in the following. The proton image, on the right of Fig. 5, displays features that appear to be complementary. Indeed, intense signal now arises from parts of the chest that consist of tissues and contain no helium gas. But the striking feature is that the lung parenchyma appears to be dark, except for a few large blood vessels (corresponding to hypointense signal in the helium image). This is partly due to the low proton density of tissues, with airspaces occupying a significant fraction of the volume, but mostly results from the very short transverse relaxation time T_2^* for protons in this highly heterogeneous tissue. Due to the magnetic susceptibility of the various components of the parenchyma, and to the small scales characterizing the alveolar structures (0.1-0.3 mm), strong internal magnetic field gradients resulting from the applied 1.5 T field very rapidly dephase the precessing magnetization, and this makes MRI of the lung a very difficult technical challenge [20]. Fortunately, these internal gradients less severely affect the precessing magnetization in the gas phase, thanks to the rapid diffusive movement of the gas atoms across alveoli that efficiently averages out spatial field variations at that scale. This motional averaging leads to NMR line narrowing, and T_2^* for the gas is long enough (>15 ms) for imaging purposes.

As was briefly mentioned in section 2.2, MRI of a low-density sample having the weak nuclear polarization (of order 10^{-5}) that results from thermal equilibrium is not feasible for SNR reasons. For instance, the ^{3}He image in Fig. 5 has been obtained with a number density of helium of order 10^{18} atoms/cc, 10^5 times lower than that of protons in water: a high, out-of-equilibrium nuclear polarization, often referred to as *hyperpolarization*, is required to compensate for this decrease in density and thus provide a high enough magnetization for MRI purposes. In this section we first describe a few striking milestones in the development of hyperpolarization of noble gases, together with the motivations that drove these developments. Some specific features of MRI with polarized gases are then discussed, and selected topic in clinical research using lung MRI with polarized helium are finally outlined.

4.1. Hyperpolarization of noble gases

In the early days of optical pumping in atomic vapors [21, 22], no method had yet been found to polarize noble gases due to the lack of convenient optical transition from the atomic ground state. Still, as early as 1956, Walters and Fairbanks used a 1-D NMR imaging technique to study a phase-separation transition in liquid isotopic helium mixtures [39]. They filled a set of 3 connected reservoirs with liquid mixtures of ^{3}He and ^{4}He (see Fig. 6, left), and obtained 3 resolved NMR lines in the applied field gradient. The line intensities in the upper and lower reservoirs were interpreted as measurements of the ^{3}He concentration (with assumptions on magnetic susceptibility), and the coexisting concentration values were plotted as a function of the temperature (Fig. 6, right). Various other methods have been used since then to obtain this phase-separation diagram with improved accuracy, but this is probably the first MRI experiment with ^{3}He, for which polarization was increased ($\times 100$) by the use of low temperatures, and it was performed more than 50 years ago.

However, it is only with the development of optical pumping methods for noble gases that almost full nuclear polarization could be achieved. Optical pumping (OP) is the redistribution of atoms among the energy sublevels of the ground

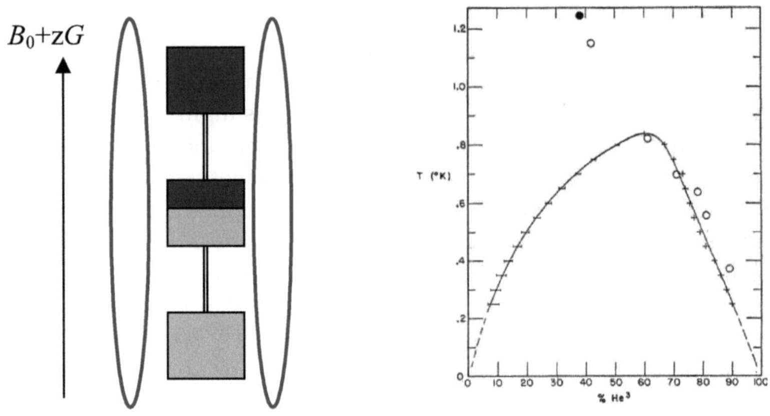

FIGURE 6. Early use of 1-D MRI in ^{3}He to study phase-separation in liquid helium mixtures. **Left:** Principle of the experiment: the interface of the phase-separated mixture lies in the central part of the cell, while the upper and lower ends contain the ^{3}He-rich and the dilute phases, respectively. The amplitudes of the frequency-resolved NMR lines from these parts of the cell are used to derive the ^{3}He concentrations **Right:** The coexisting concentrations, on the horizontal axis, are plotted for the explored temperature range (on the vertical axis), providing the first reported experimental phase-separation diagram in liquid helium mixtures (data from reference [39]).

state induced by resonant absorption of light [40]. In order to operate, OP thus requires an atomic system with (at least) two energy levels connected by an optical transition, and a light source tuned to this optical transition. The ground state (lower atomic level), and optionally the excited state (upper atomic level) have at least two sublevels, which may arise from fine or hyperfine structure terms, or from magnetic energy terms in the atomic Hamiltonian.

A very simple two-level system, used as a convenient illustrative model, is depicted in Fig. 7. Resonant light with circular polarization selectively depopulates one of the two sublevels of the ground state. Depending on the relative probabilities of spontaneous emission, a fraction of the re-emitted light takes away the angular momentum of the absorbed photon, which corresponds to inefficient OP cycles. The rest of the emitted photons have a linear (π) polarization (vertical dotted arrows in Fig. 7), and a net quantum of angular momentum is deposited in the system. Indeed, the statistical properties of an ensemble of atoms are computed using a density operator formalism to describe the populations and quantum coherences of the sublevels; in most simple cases, only atomic populations are affected by OP processes, and simple rate equations can be derived. In this frame,

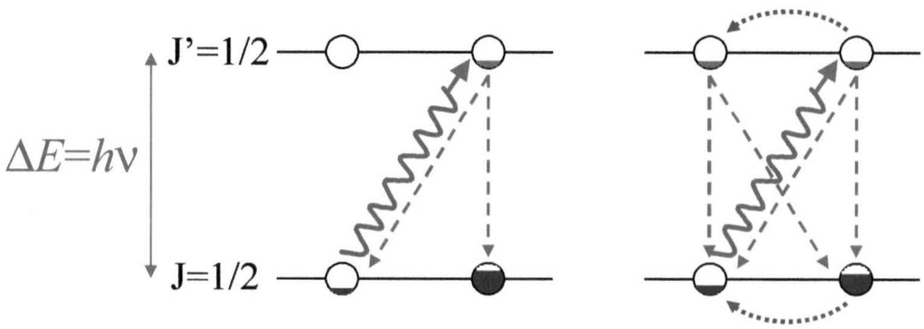

FIGURE 7. Example of OP processes in a two-level model system. The levels have angular momenta $J=J'=1/2$, and each one contains two sublevels m_J, $m_{J'}=\pm 1/2$. The sublevels are represented by circular symbols, more or less filled depending on their relative population. With a σ_+ circular polarization of the OP light tuned to the optical transition frequency ν, the transition $m_J = -1/2 \rightarrow m_{J'} = 1/2$ (wavy arrow) is selectively excited. Spontaneous emission from the excited state (downwards dotted arrows) repopulates the sublevels of the ground state. Relaxation (population transfer between sublevels) is represented by additional dotted arrows in the right-hand side diagram.

the pumping rate scales with the incident light intensity as long as saturation effects can be neglected, i.e. typically below 10mW/cm^2 to 1W/cm^{2}depending on the actual system.

When relaxation processes (additional arrows in the right hand side part of Fig. 7) efficiently transfer atoms between the sublevels of a given state, the overall OP efficiency can be significantly altered. Relaxation usually results from atomic collisions in the gas, and its rate strongly depends on the atomic state symmetry. For instance, even at moderate density (e.g. 10^{17}cm^{-3}, or a few mbar at room temperature), relaxation rates can be higher than the radiative decay rates for a P or D state, but they are negligible for an S state. Assuming full redistribution of the atomic populations among the excited sublevels, the emitted light is unpolarized and the OP efficiency fully results from a depopulation mechanism, that can still be very efficient. In contrast, any relaxation in the ground state tends to reduce the OP efficiency; OP will thus easily and efficiently operate only if the atomic ground state is weakly affected by collisions, or at very low atomic densities (e.g. in an atomic beam).

As already mentioned, no convenient optical transition allows to perform OP on the ground state of any of the noble gases; among other reasons, they lie in the far-UV range. This is unfortunate since among their various isotopes, some of these atoms have a well-shielded, purely nuclear angular momentum. Especially for $I=1/2$, i.e. for ^{3}He and ^{129}Xe, very long relaxation times (hours) of the nuclear polarization can be obtained in spite of frequent atomic collisions with other atoms or with adequately chosen cell walls. However, two indirect OP methods have been discovered in the 1960's to override this problem.

The first one, spin-exchange OP (SEOP) was demonstrated in 1960 [41], but at that time it only provided very low nuclear polarizations (0.01%). Following this early demonstration, the reasons of its limited performance have been systematically studied, understood, and bypassed [42]. SEOP consists in performing OP on an alkali vapor mixed with a noble gas in a cell, usually Rb (see Fig. 8). Due to the

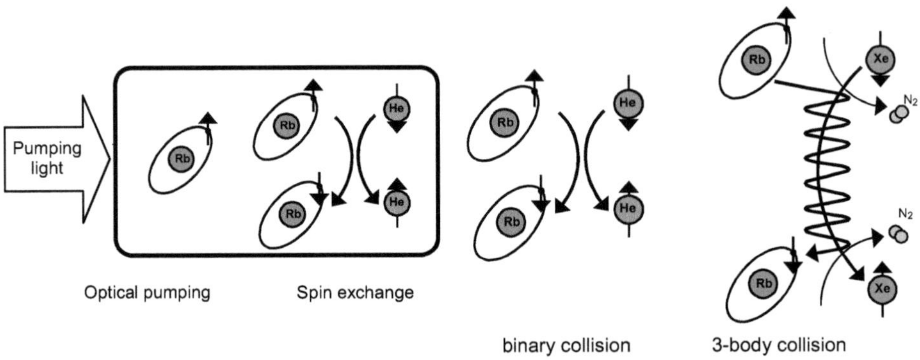

FIGURE 8. Spin-exchange OP in a Rb - ^{3}He or ^{129}Xe mixture; OP is performed on the alkali atom, and its electronic orientation creates nuclear orientation of the noble gas during binary collisions (the dominant process for ^{3}He SEOP) or 3-body collisions (the dominant process for ^{129}Xe SEOP).

complicated structure of the atomic levels involved in the OP process (both Rb isotopes have a non-zero nuclear spin giving rise to hyperfine-structure sublevels), the OP process is not as simple as depicted in Fig. 7. Moreover, radiative decay from the excited state is advantageously quenched by collisions with added N_2 gas to prevent reabsorption of unpolarized light. Still, using suitable powerful lasers for the 795 nm transition of Rb, an efficient depopulation OP process allows to sustain a high degree of electronic polarization for the Rb atoms, that is transferred to the nuclei noble gas atoms during collisions via transient hyperfine coupling.

The second indirect method, metastability exchange OP (MEOP), was discovered in 1963 [43] and exclusively applies to ^{3}He. It is interesting to note that in this first experiment, NMR measurements were performed and the optical detection of pulsed NMR was performed together with standard cw NMR (Fig. 9). This attention paid to NMR is probably due to the fact that this work was led by G.K. Walters, who conducted the NMR experiment depicted in Fig. 6 a few years before, and to the interest in magnetometry that motivated this research.

In a helium gas, MEOP is actually performed between two excited levels, the lower 2^3S level being metastable with a radiative lifetime of several thousands seconds (see Fig. 10). This level, which acts as a ground state for the OP process, is

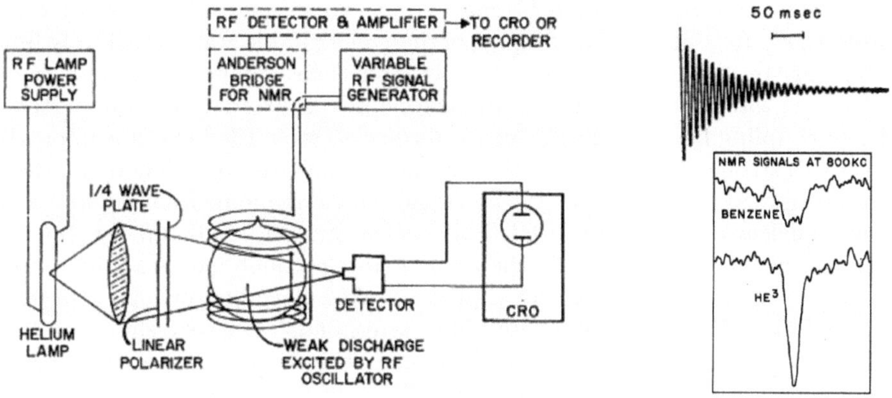

FIGURE 9. Sketch of the first MEOP experiment providing high nuclear polarization in ^{3}He. Optical detection of NMR precession (top right) provided a much higher SNR than cw NMR (bottom right). Figures are from reference [43].

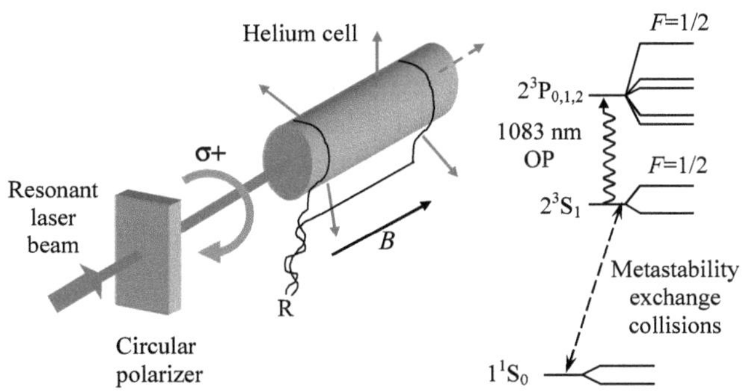

FIGURE 10. Left: Schematic view of a MEOP setup. In a low-pressure helium-3 gas cell, a weak RF discharge promotes a small fraction ($\sim 10^{-6}$) of the atoms into the excited metastable state 2^3S, where resonant absorption of the circularly polarized 1083 nm light can occur. Nuclear polarization is transferred to the ground state atoms by metastability exchange collisions. Long OP cells can be used thanks to the weak absorption of OP light, especially by a polarized gas. Right: Atomic levels of ^{3}He and physical processes involved in the OP cycle. The 2^3S and 2^3P states are composed of 2 and 5 sublevels, respectively, and the 1083 nm line has a complex structure. An efficient OP transition actually connects the highest-lying $F=1/2$ sublevels of the 2^3S and 2^3P states, and the OP process is similar to that sketched in Fig. 7.

usually populated by electron collisions in a plasma discharge in a gas at moderate density (10^{16}-10^{17} cm^{-3}). The lifetime of these metastable states is actually limited to 1-10 ms by diffusion to the cell walls, and their number density is of order 10^{-6} of the total density. For the ^{3}He isotope in this excited metastable state, an efficient coupling between the nucleus and the electrons (the hyperfine interaction) results in a strong entanglement of electronic and nuclear spins. Therefore the OP-enforced optical orientation of the electronic angular momentum simultaneously induces nuclear orientation as well. This nuclear orientation is rapidly transferred to the atoms having remained in the ground state through metastability exchange collisions. This important collisional process takes place between the true ground state and the 2^3S metastable state, and corresponds to a very short interaction resulting in a fast exchange of the electronic excitations of the colliding atoms, with no change in the nuclear orientations.

Each of the two methods has its own advantages and limitations. For instance, SEOP operates both for ^{3}He and ^{129}Xe, and is used for numerous studies involving polarized gases [44], e.g. in precise nuclear co-magnetometers in the search of the electric dipole moment of neutrons [45]. Progress in laser technology at 795 nm, at first with Ti:Sapphire lasers, then with dedicated diode laser arrays, has made it an efficient tool for all applications requiring high-density polarized gases, such as spin filters for cold neutrons [46] or lung imaging [47]. To date, it is the only method to polarize xenon, of which large amounts can now be obtained with a high polarization (e.g. 0.3 liter/hour with 50% polarization [48]). For ^{3}He, SEOP directly operates at high pressure (several bars), but is a slow process (several hours are required to obtain a sizable polarization). This difference originates from the much lower spin-exchange cross section that ensures polarization transfer from Rb to He., and current studies using mixtures of alkali atoms yield promising results [49].

When laser light with adequate spectral characteristics is used, MEOP provides very high nuclear polarization (>70%) with good efficiency (>1 polarised nucleus per absorbed photon) [50]. With the recent development of spectrally suitable high power fiber lasers [51], the only drawback of this method is its limited range of operating pressures, of order 0.5-5 mbar, required to have a suitable plasma discharge in which metastable atoms are populated [52, 53]. Whenever a higher final pressure is needed, non-relaxing compression of the gas is required. This is the case for several applications including lung MRI, and it introduces demanding requirements on the gas handling and compressing devices. However, it was recently shown that the range of operating pressures can be extended to several tens of mbar by performing MEOP in a high magnetic field, such as the 1.5 T field of an MRI system [54, 55], which could make subsequent compression less difficult to perform.

The first attempt at compressing ^{3}He after it has been polarized by MEOP dates back to 1968 [56], using cryogenic means to increase the gas density and ultimately liquefy it. But only limited polarization (of order 1%) could be obtained due to fast wall relaxation. It was only with the use of non-relaxing cryogenic coatings providing long T_1s [57] that polarized ^{3}He could be successfully cooled as a highly polarized gas, or liquid [58], allowing for instance studies of the effect of spin-polarization on quantum properties of gaseous ^{3}He [59, 60]. This series of studies in dense polarized systems obtained by cryogenic methods is currently focused on the non-linear NMR behavior that results from the action of distant dipolar fields in hyperpolarized liquids [61].

The following attempt at compressing polarized ^{3}He was made in 1970 using a mechanical Toepler pump (a mercury piston moving in a glass vessel) and only met with a limited success (less than 4% polarization at 0.3 bar) [62] due to the lack of powerful OP light sources. Mechanical compression attempts resumed only after the first multi-watt lasers for MEOP have been developed in the late 1980s [63], this time succeeding in reaching sufficient polarization and pressure for use in various experiments [64, 65], in particular to prepare polarized ^{3}He targets for the measurement of the neutron electric form factor in scattering experiments of polarized electron beams [66]. Since then, various non-relaxing mechanical compression techniques have been applied to polarized ^{3}He, using a diaphragm pump [67], a peristaltic pump [68], or a piston compressor [69, 70] for applications in lung MRI, in more accurate electron scattering experiments, and in neutron spin filters [71, 72].

A convenient feature of MEOP-based polarization systems is the possibility to up- or down-scale the apparatus depending on requirements, and on available budget. Commercially available fiber lasers[9] deliver up to 15 W. The optimized large-scale system currently in operation at Mainz University, that is used to produce polarized gas for lung MRI, for scattering experiments and for neutron spin filters, makes use of five OP cells, each one being over 2 m-long, to efficiently absorb as much as possible of the available pumping power [70]. With the high quantum efficiency of MEOP, this system routinely yields 1.2 bar$\times$liter/hour of 80%-polarized ^{3}He, or 3.3 bar$\times$litre/hour of 60%-polarized ^{3}He [70]. To date, systems with similar performance operate only at the ILL in Grenoble [69] and in Garching[10] for neutron spin filters. For all applications, polarized ^{3}He gas must be shipped to the final user since these systems are too bulky and delicate to be transported. Storage cells suitable for transportation, i.e. with very long relaxation times and magnetic shielding to avoid loss of polarization, have been developed, and shipping of hundreds of liters of polarized gas has now been successfully performed [70]. More compact systems, in which cell size, laser power, and compressor flow rate are all consistently lower, have of course a significantly reduced performance, yielding e.g. 0.2-0.3 bar$\times$liter/hour of 50%-polarized ^{3}He [67, 68]. Still, for less demanding applications such as methodological developments in MRI, lung

[9] We are aware of two companies having provided lasers at 1083 nm with the 2-GHz bandwidth matched to the Doppler absorption profile of He atoms: Keopsys (Lannion, France) and IPG Photonics (Burbach, Germany).

[10] The gas polarizer used in Garching was manufactured by IC-Automation, a Mainz-based company who built this sole system using the expertise of the research team of Mainz University. They have currently no plans to make more gas polarizing systems.

imaging in small animals, or imaging with a limited number of patients, these systems might offer an increased ease of use by providing locally-polarized gas on demand.

These options to gas provision must indeed be compared to the use of SEOP-based systems. In-house SEOP systems have been developed by several groups, and in addition a dozen commercial units[11] have been implemented in selected hospitals for research on lung MRI. These systems routinely yield 1 liter of 30% polarized gas after 8-12 hours of OP, which is currently less than what both kinds of MEOP-based systems can provide.

4.2. Specific features of imaging with polarized gases

The most obvious difference between MRI with polarized noble gases and usual proton MRI is of course the lower operating frequency which results from the lower value of the gyromagnetic ratio (30% lower for ^{3}He, 3.6 times lower for ^{129}Xe). All rf systems, including transmit and receive coils, thus have to be designed for these unusual frequencies. However, more than ten years after the first lung image was obtained using prototype systems, commercial solutions now exist and operation at these frequencies is now a standard option for many systems.

The main difference that makes MRI with polarized gases so unusual for untrained MR physicists or radiologists is the non-renewable character of the polarization. Nuclear relaxation has to be considered as a plague, not as a natural way to obtain signals after just waiting a few seconds (the usual T_1 value for protons in tissues). Gas management has to carefully avoid relaxation losses at all stages, from the preparation of the polarized gas until the end of the image acquisition. Magnetic relaxation induced by walls of gas containers is now well understood and controlled [73, 74, 75], and all materials in contact with the gas (tubes, valves, mouthpieces or respiratory masks...) must be selected with care. Magnetic relaxation can also occur in the bulk of the gas due to atomic diffusion in inhomogeneous magnetic field, with a typical rate:

$$1/T_1^{\mathrm{magn.}} \sim D(G/B)^2, \tag{12}$$

that scales with the diffusion coefficient D (2 cm^2/s at 1 bar for ^{3}He) and with the square of the relative field inhomogeneity (G/B in the case of a uniform gradient G). This shows in particular that polarized gas should not dwell in parts of the MRI magnet where the field has a steep variation with position. A last important source of relaxation lies in collisions with the paramagnetic O_2 molecules of air, which induce a decay rate:

$$1/T_1^{\mathrm{magn.}} \sim 0.41 \times pO_2, \tag{13}$$

[11] The polarizers have been developed on the late 1990s by MITI, Magnetic Imaging Technologies, Inc., a spinoff company from the Princeton Group that was involved in the first lung imaging demonstration [36]. MITI was then sold to Nycomed-Amersham, which later became Amersham Health, which in turn was later sold to GE Healthcare. The polarizers are still maintained in operation, but their future status is unknown.

where pO_2 stands for the partial pressure of O_2 (in bars). This value is inferred from reference [76] for room temperature, and corresponds to a relaxation time of 11.4 s in room air (with 21% O_2). Polarized ^{3}He must thus be preserved from exposure to air until the time at which it is inhaled. Polarization decays rather rapidly in-vivo, and data acquisition must be performed in a matter of seconds - which anyhow usually corresponds the maximum time patients with respiratory problems can hold their breaths. This relaxation process is indeed a constraint, but can also be used to evaluate the O_2 contents of the lungs (see Fig. 14 in the next section). The last feature of imaging with a hyperpolarized sample is that the sequence of rf pulses must be tailored for an optimal use of the available magnetization, either by an appropriate choice of successive small tip angles, or by using multiple echo acquisition whenever possible. Since polarized gas is not easily available, and so far only in rather limited amounts, a careful planning of the patients' imaging sessions is necessary.

A second specific series of features of MRI with polarized gases originates from the fast diffusion of atoms in a gas phase. The beneficial consequence on transverse relaxation induced by internal gradients has already been discussed at the beginning of section 4, but adverse effect of diffusion also exist. For instance, the spatial resolution is intrinsically limited by diffusion: for a given amplitude G of an imaging gradient, the spatial resolution is $\delta x = 1/\gamma G T_{\mathrm{obs}}$ (see section 3.1), while diffusion-induced attenuation limits the useful time to $T_{\mathrm{obs}} \sim D\delta x^2$. The maximum resolution thus scales as $\delta x \sim (D/\gamma G)^{1/3}$. For standard MRI systems, for which G cannot exceed 30 mT/cm, the highest resolution for ^{3}He diffusing in air is of order 0.4 mm, not quite small enough to directly image alveolar structures: MR microscopy is not possible with gases. However diffusion weighting in MRI offers the potential to probe gas diffusion over a wide range of time and paces scales, linked by the same relation that directly limits image resolution: $T_{\mathrm{obs}} \sim D\delta x^2$. Characterizing apparent diffusion at various scales within the lung may indirectly provides relevant information on the lung microstructure.

A last specific feature of NMR or MRI with polarized samples is that the SNR is field-independent over a wide range of field values B_0 (see section 2.2). The direct influence of B_0 is limited to the effect of internal gradients on the signal characteristics. From this point of view, high fields (1.5 T and above) induce short values for T_2^*, that may be more sensitive to the alveolar characteristic size and shape, thus providing a physiologically relevant contrast mechanism [77]. Conversely, transverse relaxation times are longer at reduced fields, as demonstrated for instance at 0.1 T [78] and at 3 mT [79]. Weaker field gradients can then be used for imaging or for ADC measurements. This actually improves SNR in images and allows measuring diffusion coefficients at longer time scales, thus providing more information on the connectivity of lung air spaces. Monitoring the decay of spin-echo trains provides a precise way of measuring ADC coefficients in this case. At ultra-low fields, internal gradients are so low that the observed decay in spin-echo experiments results from the oxygen-induced relaxation process for which Eq. 13 also applies [79].

Another clear advantage of operating at low field is the reduced cost and increased flexibility of the imaging system: an open geometry can be designed, and standing or sitting patients may be examined. Initial studies at ultra low field in home made vertical scanners have demonstrated the potential of the technique [80, 81]. This could open up possibilities of low-cost dedicated scanners for the screening or follow-up of lung diseases.

4.3. Current status of research

In this section we present a selection of results chosen to illustrate the potential of lung MRI in humans with polarized ^{3}He, deliberately overlooking all the important work done using ^{129}Xe or in animal models, for which much information can be found in the literature (e.g. [82]). These results have been obtained by clinical research teams participating in the PHIL project (Polarized Helium to Image the Lung), a joint effort of nine European research teams in five countries [83]. An important objective of the project was to demonstrate the potential and the validity of the new MRI method as a diagnostic and prognostic tool for given lung pathologies: emphysema and selected Chronical Obstructive Pulmonary Diseases (COPD), such as bronchitis and bronchiolitis. This choice was motivated by the frequent occurrence of these diseases and the very high cost of their treatment for society: 10% of the population and 25% of the smokers suffer from COPD, which is the fourth cause of mortality in Europe and in the USA. The core of the project was to perform a clinical trial on a large group of patients with the ^{3}He MRI method and with conventional techniques: pulmonary function tests, High Resolution Computed Tomography (HRCT), Krypton scintigraphy. An important objective of the PHIL project was to provide new tools for the study of COPD, aiming at differentiation of various types of diseases, as well as their detection at an early stage, with expectation that in the long range the findings of the project could lead to monitoring therapeutic treatment. The clinical trial successfully enrolled a total of 116 subjects (62 COPD, 17 alpha-1-antitrypsin deficiency (ATD) and 37 healthy volunteers). The collaborative work has triggered the dissemination of the method in Europe. In addition to the trials performed in Mainz, Sheffield and Copenhagen, animal model studies have been made in Lyon and Madrid, and methodological developments in Paris area, Cracow and Mainz. The collaboration is currently pursued, including several new participants, in the Phelinet project that aims at promoting the training of young researchers and scientific exchanges between participants [84].

An important objective of research with ^{3}He MRI is the systematic comparison of MR images with other existing imaging modalities. Sets of images obtained by ^{3}He ventilation images obtained during 12 s-long breath holds and by ^{81m}Kr single photon emission computed tomography (SPECT)[12] are displayed in

[12] A cyclotron produces the radioactive isotope ^{81}Rb, that decays to metastable ^{81m}Kr, which the patient inhales. The metastatable isotope decays and emits gamma photons that are detected. A

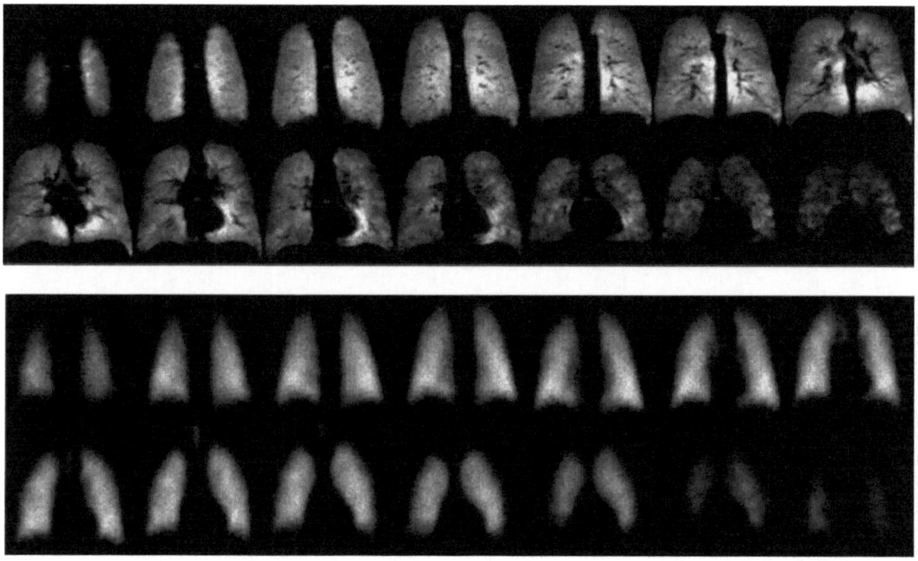

FIGURE 11. Lung images in a normal volunteer: ^{3}He MRI (top), and ^{81m}Kr SPECT (bottom).- Coronal slices from the posterior part of the lung to the anterior part, from the upper left corner. *Courtesy of Trine Stavngaard and Lise Vejby Soegaard, Danish Research Centre for Magnetic Resonance, Copenhagen University Hospital, Hvidovre, Denmark.*

Figs. 11 (for a normal volunteer) and 12 (for a COPD patient). In both sets of images, the spatial resolution provided by ^{3}He MRI is far superior (1.3×1.3 mm^2 in each 10 mm-thick slice), and much more detailed information is provided on ventilation defects in Fig 12. The study included 32 patients, and showed good correlation between the ventilation defects assessed by the two methods, in spite of the difference in lung inflation conditions for the two methods (17 min.-long SPECT images were acquired close to tidal volume during normal breathing) [85].

Much higher time-resolution can be obtained using a sliding window radial acquisition scheme [86]. Projection images obtained by this method are shown in Fig. 13, with a time interval between displayed images of 0.54 s. For the normal volunteer, the temporal passage of gas down the trachea, into the bronchi and peripheral lung is clearly resolved. For the COPD patient, the image series shows regions of ventilation obstruction in both lungs, particularly in the upper lobes,

gamma camera records planar projections (scintigrams) showing the ventilated area in the lung. From several scintigrams taken from various angles, a 3-D reconstruction of the lung ventilation is obtained. For the images in Figs 11 and 12, the spatial resolution was 8 mm, the radiation dose was 1/2 mSv, and the imaging time was 17 min.

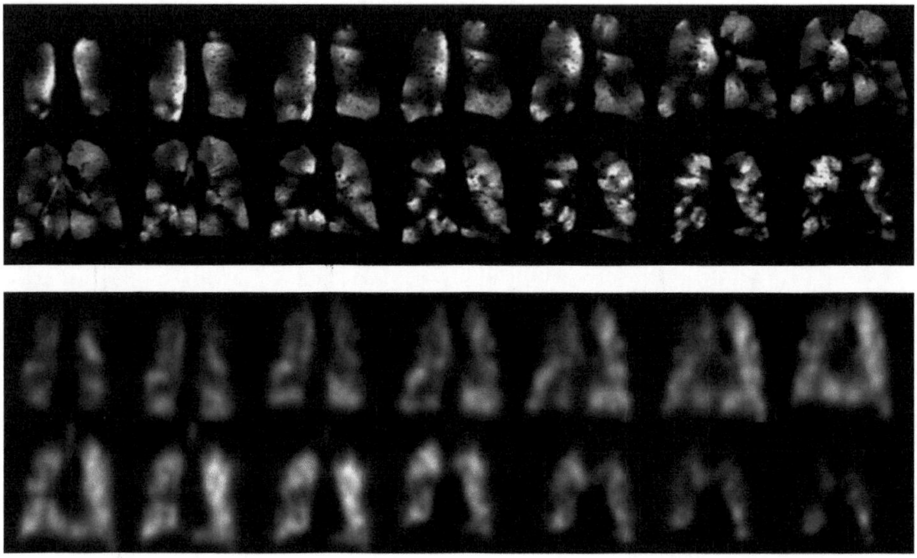

FIGURE 12. Same ^{3}He MRI and ^{81m}Kr SPECT lung images as in Fig. 11 for a COPD patient. *Courtesy of Trine Stavngaard and Lise Vejby Soegaard.*

and a delayed emptying/depolarization of gas in the lower left lobe which could be indicative of air trapping. Such dynamic ventilation maps may provide valuable information on ventilation defects in COPD patients, complementary to the static ventilation defect data obtained during a breath hold (Figs. 11 and 12) [87].

Decay of signal during a breath hold resulting from O_2-induced relaxation (Eq. 13) can be recorded in a series of images, from which maps of the local partial pressure of oxygen are computed, as displayed in Fig. 14 for a normal subject and a COPD patient [88]. When low-resolution images are acquired, pO_2 values are determined with such high accuracy that their time-evolution provides information on regional oxygen uptake. It is dependent on lung perfusion, and thus the uptake map can be seen as a regional ventilation perfusion (V_A/Q) map of the lung [89], allowing a direct measurement of the regional V_A/Q distribution in a noninvasive fashion at relatively high resolution when compared to ventilation perfusion scintigraphy.

Figure 15 displays examples of ADC maps obtained using a bipolar diffusion-sensitization gradient, a standard technique which provides information on diffusion of ^{3}He atoms over relatively short times and distances. Such images and the related histograms of increased ADC values have been shown to correlate well with enlarged alveolar sizes of patients with emphysema [90].

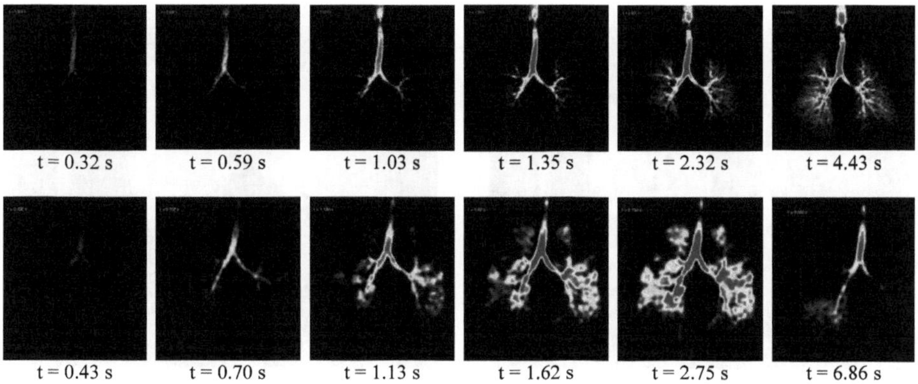

FIGURE 13. Dynamic projection images obtained using a radial sequence (time resolution: 0.135 s). **Top:** selected images in a series from a healthy normal subject recorded during an inhalation of 300 ml of 40%-polarized ^{3}He gas. **Bottom:** Similar dynamic time series from a COPD patient. *Courtesy of Jim Wild, Academic Radiology Dept., U. of Sheffield, UK.*

These few sets of images have been chosen as illustrative examples of the various relevant physiological information that MRI with polarized ^{3}He may quantitatively provide. Similar research has been indeed performed since the early days of polarized gas MRI also in the USA, where the University of Virginia has been, and remains, one of the most active research centers. It is pursued by tens of teams worldwide, including in Canada and in Japan, who are involved in lung MRI with polarized ^{3}He or ^{129}Xe. More than 1000 patients have now been subject to various imaging protocols, and the potential of MRI in the early detection and in the staging of disorders such as asthma, emphysema, and cystic fibrosis has been explored. Its potential interest has also been suggested for very different problems such as the follow-up of lung transplants, or the comparison of lung development during childhood for normal children [91] and for children with a pre-term birth history [92].

5. Conclusion and prospects

Looking back at the history of NMR and of MRI, one is impressed by its fast development as a research field, as a tool, as an industrial opportunity, and as widespread clinical imaging modality. NMR moved rapidly from fundamental research at Stanford University to commercialization because of Varian's involvement. From that point, industry led the development, that depended more on technology than on fundamental research. Later on, through most of MRI's evolution, the challenges were to improve image quality and to reduce scan times, which

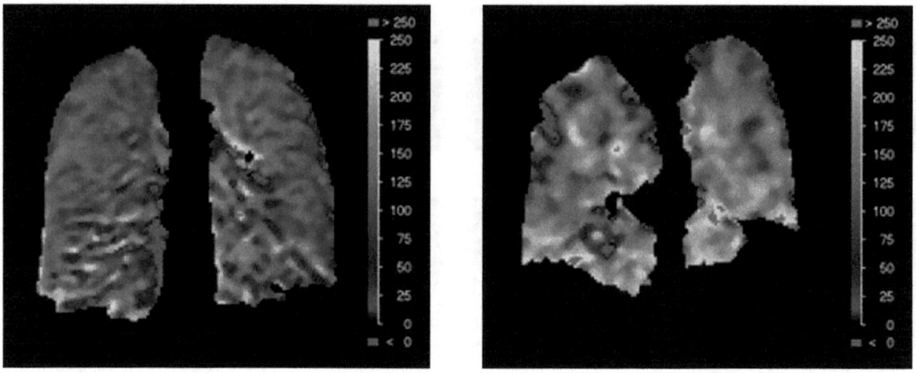

FIGURE 14. Measurement of oxygen concentration in the lung using ^{3}He imaging. **Left:** Image of a healthy volunteer with relatively homogeneous distribution of oxygen. **Right:** Image of a patient with chronic obstructive pulmonary disease (COPD) demonstrating inhomogeneous distribution of oxygen because of regional impairment of oxygen uptake by blood. Thus, alveolar oxygen concentration remains elevated in these regions when compared with regions with normal ventilation and perfusion. *Data from reference* [88] - *Courtesy of Wolfgang Schreiber, Johannes Gutenberg-University Mainz, Medical School, Germany.*

again involved a lot of technology. As regards MRI, every decade was marked by spectacular qualitative and quantitative progress. Following the physicists' early demonstrations in the 70s, leading instrumentation companies heavily invested in research and development in the 80s, and started clinical evaluation of images. The 90s were the time of the increase in available equipment for routine use, and of the growing clinical usefulness of the images. In the recent years, together with a steady increase in the number of operational imaging sites, there is an impressive reduction of scan times with the use of revolutionary methods for data acquisition and signal processing.

It can be argued that one key factor for this success story probably lies in its excellent timing [93]. For instance, two decades earlier, the technology for whole-body superconducting magnets and, more importantly, for computers performing all the necessary Fourier transforms, simply did not exist. Two decades later, the demand for diagnostic imaging would have been less compelling, and better-established imaging modalities using well-controlled ionizing radiations (e.g. CT-scans) would possibly have been considered as sufficient golden standards. Both the equipment industry and the radiologic profession would probably have required a strong incentive to accept such innovative methods and techniques. But even more important is the fact that the regulatory environment has drastically changed since the 80s. With the increasing concern regarding the impact of static and

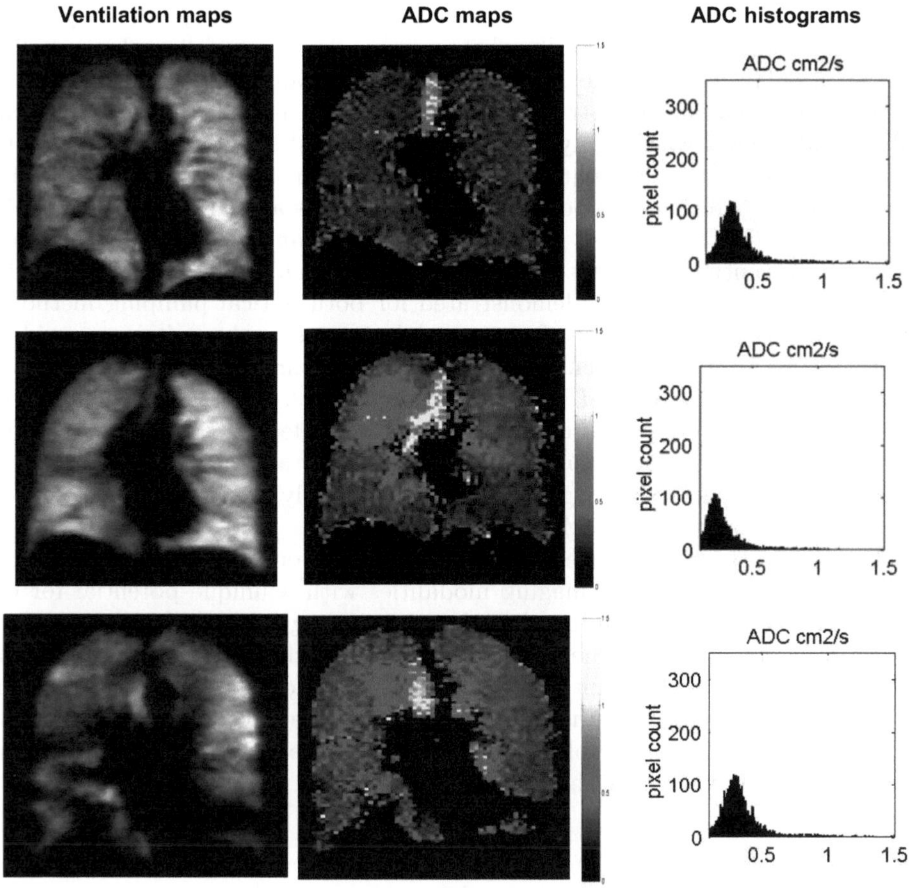

FIGURE 15. Left to right: ^{3}He MRI gas density projection images, ADC maps and ADC histograms. Top to bottom: data from a 48-year-old female healthy non-smoker, a 51-year-old female healthy smoker, and a 62-year-old male diagnosed COPD patient. *Courtesy of J. Wild, Academic Radiology Dept., U. of Sheffield, UK.*

oscillatory fields, the level of proof of safety required for new instruments based on new principles is now tremendous. As a result, the development costs until the obtention of FDA approval in the USA, and of the equivalent certification in other countries, has become so high that it would probably be impossible to fund, neither by industry, who now have limited freedom for such ventures, nor by academic institutions, who have increasingly limited budget for unscheduled research.

The history of the new MRI technique for imaging the lung - the only organ that conventional MRI can hardly observe - has in a way followed a parallel, or rather a delayed path. Like NMR, it appeared as an unexpected outcome of decades of research and of instrumental developments aiming at totally different fundamental studies with highly polarized helium-3. With optical pumping, the stage was set since the 60s, but the main character in the story - the powerful laser needed for sufficient gas polarization - delayed its appearance until the 90s. Up to now, access to polarized gas has been difficult and this remains the main bottleneck preventing many research groups to become active in the field. The technology exists, it has been demonstrated for both optical pumping methods, including with a small number of commercial systems, and yet it is not widely developed. The arguments given by potential manufacturers include the uncertain status of patents regarding MRI with polarized gases, the increasing difficulty and cost of certification processes: since gases are inhaled, they are considered as drugs, and approval is consequently much harder to obtain than for a new type of coil or of MR sequence. Such arguments are disturbingly reminiscent of the above uchronic discussion on proton MRI.

In spite of these differences in timing, both proton MRI and polarized gas MRI are incredibly versatile imaging modalities with a unique potential for diagnosis. They are totally non-invasive, they can be rugged tools for the clinician or powerful instruments for the most advanced research - one may think, for instance, of the studies of cognitive processes using functional MRI of the brain. The remaining challenges for the dissemination and use of helium-3 MRI are now probably off the hands of the physicists who contributed to the beginning of this story, but they all wish it will have a happy end.

References

[1] I.I. Rabi, S. Millman, P. Kusch, and J. R. Zacharias, *A new method of measuring nuclear magnetic moments*, Phys. Rev. **53** (1938), 318; I.I. Rabi, S. Millman, P. Kusch, and J. R. Zacharias, *The molecular beam resonance method for measuring nuclear magnetic moments. The magnetic moments of Li6, Li7 and F19*, Phys.Rev. **55** (1939), 526.

[2] E.M. Purcell, H.C. Torrey, and R.V. Pound, *Resonance absorption by nuclear magnetic moments in a solid*, Phys. Rev. **69** (1946), 37.

[3] F. Bloch, W. W. Hansen, and M. Packard, *Nuclear induction* Phys. Rev. **69** (1946), 127; F. Bloch, W. W. Hansen, and M. Packard, *The nuclear induction experiment*, Phys. Rev. **70** (1946), 474.

[4] F. Bloch, *Nuclear induction*, Phys. Rev. **70** (1946), 460.

[5] H.C. Torrey, *Transient nutations in nuclear magnetic resonance*, Phys. Rev. **76** (1949), 1059.

[6] E.L. Hahn, *Nuclear induction due to free Larmor precession*, Phys. Rev. **77** (1950), 297.

[7] E.L. Hahn, *Spin echoes*, Phys. Rev. **80** (1950), 580.

[8] J.T. Arnold, S.S. Dharmatti, and M.E. Packard, *Chemical effects on nuclear induction signals from organic compounds*, J. Chem. Phys. **19** (1951), 507.

[9] C.J. Gorter and L.J.F. Broer, *Negative result of an attempt to observe nuclear magnetic resonance in solids*, Physica **9** (1942), 591.

[10] R.J. Singer, *Blood-flow rates by NMR measurements*, Science **130** (1959), 1652.

[11] J.A. Jackson and W.H. Langham, *Whole-body NMR spectrometer*, Rev. Sci. Instrum. **39** (1968), 510.

[12] P.C. Lauterbur, *Image formation by induced local interactions: examples of employing nuclear magnetic resonance*, Nature **242** (1973), 190.

[13] A.N. Garroway, P.K. Grannell, and P. Mansfield, *Image formation in NMR by a selective irradiative process*, J. Phys. C **7** (1974), L457.

[14] P. Mansfield and A.A. Maudsley, *Planar spin imaging by NMR*, J. Magn. Reson. **27** (1977), 101.

[15] R.R. Ernst and W.A. Anderson, *Application of Fourier transform spectroscopy to magnetic resonance*, Rev. Sci. Instrum. **37** (1966), 93.

[16] A. Kumar, D. Welti, and R.R. Ernst, *NMR-Fourier-Zeugmatography*, J. Magn. Reson. **18** (1975), 69.

[17] W.A. Edelstein, J.M.S. Hutchison, G. Johnson and T.W. Redpath, *Spin warp NMR imaging and applications to human whole-body imaging*, Phys. Med. Biol. **25** (1980), 751.

[18] From the MEDLINE database of the U.S. National Library of Medicine. 4600 international journals in all areas of the life sciences, with particular emphasis on biomedicine are indexed.

[19] S. Gleyzes, et al., *Quantum jumps of light recording the birth and death of a photon in a cavity*, Nature **446** (2007), 297.

[20] E.D. Pracht, J.F.T Arnold, T.T. Wang, et al., *Oxygen-enhanced proton imaging of the human lung using T2**, Magn. Reson. Med. **53** (2005), 1193.

[21] A. Kastler, *Optical methods of atomic orientation and of magnetic resonance*, J. Opt. Soc. Am. **47** (1957), 460.

[22] B. Cagnac, J. Brossel and A. Kastler, *RMN du mercure Hg-201 aligné par pompage optique*, C.R Acad. Sci. **246** (1958), 1827.

[23] J. Jeener, *Equivalence between the "classical" and the "Warren" approaches for the effects of long range dipolar couplings in liquid NMR*, J. Chem. Phys. **112** (2000), 5091.

[24] D.I. Hoult and P.C. Lauterbur, *The sensitivity of the zeugmatographic experiment involving human samples*, J. Magn. Reson. **34** (1979), 425.

[25] J. Bittoun, B. Querleux and L. Darrasse, *Advances in MR imaging of the skin*, NMR Biomed. **19** (2006), 723; L. Darrasse and J.-C. Ginefri, *Perspectives with cryogenic RF probes in biomedical MRI*, Biochimie **85** (2003), 915.

[26] M. Mossle, et al., *SQUID-detected in vivo MRI at microtesla magnetic fields*, IEEE Transactions on Applied Superconductivity **15** (2005), 757.

[27] D. Budker and M. Romalis, *Optical magnetometry*, Nature Physics **3** (2007), 227.

[28] H.J. Mamin, M. Poggio, C.L. Degen and D. Rugar, *Nuclear magnetic resonance imaging with 90-nm resolution*, Nature Nanotechnology **2** (2007), 301.

[29] J. Pauly, P. Le Roux, D. Nishimura and A. Macovski, *Parameter relations for the Shinnar-Le Roux selective excitation pulse design algorithm*, IEEE Trans. Med. Imaging **10** (1991), 53.

[30] G. McGibney, M.R. Smith, S.T. Nichols and A. Crawley, *Quantitative evaluation of several partial Fourier reconstruction algorithms used in MRI*, Magn. Reson. Med. **30** (1993), 5159.

[31] D.K. Sodickson and W.J. Manning, *Simultaneous acquisition of spatial harmonics (SMASH): Fast imaging with radiofrequency coil arrays*, Magn. Reson. Med. **38** (1997), 1603.

[32] K.P. Pruessmann, M. Weiger, M.B. Scheidegger and P. Boesiger, *SENSE: Sensitivity encoding for fast MRI*, Magn. Reson. Med. **42** (1999), 2962.

[33] M. Lustig, D. Donoho and J.M. Pauly, *Sparse MRI: The application of compressed sensing for rapid MR imaging*, Magn. Reson. Med., in press, DOI: 10.1002/mrm.21391 (2007).

[34] M. Goldman, H. Johannesson, O. Axelsson and M. Karlsson, *Hyperpolarization of 13C through order transfer from parahydrogen: a new contrast agent for MRI*, Magn. Reson. Imaging. **23** (2005), 153.

[35] S. Mansson, et al., *13C imaging-a new diagnostic platform*, Eur Radiol. **16** (2006), 57.

[36] M.S. Albert, G.D. Cates, B. Driehuys, et al., *Biological MRI using laser-polarized 129Xe*, Nature **370** (1994), 199.

[37] J.R. MacFall, H.C. Charles, R.D. Black, et al., *Human lung air spaces: potential for MRI with hyperpolarized 3He*, Radiology **200** (1996), 553.

[38] H.U. Kauczor, D. Hofmann, K.F. Kreitner, et al., *Normal and abnormal pulmonary ventilation: Visualization at hyperpolarized 3He MRI*, Radiology **201** (1996), 564.

[39] G.K. Walters and W.M. Fairbank, *Phase separation in 3He-4He solutions*, Phys. Rev. **103** (1956), 262.

[40] A. Kastler, *Méthodes optiques d'étude de la résonance magnétique*, Physica, **17** (1951), 191.

[41] M.A. Bouchiat, T.R. Carver and C.M. Varnum, *Nuclear Polarization in He3 Gas Induced by Optical Pumping and Dipolar Exchange*, Phys. Rev. Lett. **5** (1960), 373.

[42] W. Happer, *Optical Pumping*, Rev. Mod. Phys. **44** (1972), 169.

[43] F.D. Colegrove, L.D. Schearer and G.K. Walters, *Polarization of He3 gas by optical pumping*, Phys. Rev. **132** (1963), 2561.

[44] T.G. Walker and W. Happer, *Spin-exchange optical pumping of noble-gas nuclei*, Rev. Mod. Phys. **69** (1997), 629.

[45] D. Bear, et al., *Limit on Lorentz and CPT violation of the neutron using a two-species noble-gas maser*, Phys. Rev. Lett. **85** (2000), 5038 - erratum in *Phys. Rev. Lett.* **89**, 209902.

[46] G.L. Jones, et al., *Test of He-3-based neutron polarizers at NIST*, Nuc. Instr. Meth. A **440** (2000), 772.

[47] T. Chupp and S. Swanson, Adv. At. Mol. Opt. Phys. **45** (2001), 51.

[48] I.C. Ruset, S. Ketel and F.W. Hersman, *Optical pumping system design for large production of hyperpolarized 129Xe*, Phys. Rev. Lett. **96** (2006), 053002.

[49] W.C. Chen, T.R. Gentile, T.G. Walker, et al., *Spin-exchange optical pumping of He-3 with Rb-K mixtures and pure K*, Phys. Rev. A **75** (2007), 013416.

[50] P.J. Nacher and M. Leduc, *Optical pumping in 3He with a laser*, J. Phys. (Paris) **46** (1985), 2057.

[51] G. Tastevin, S. Grot, E. Courtade, S. Bordais and P.-J. Nacher, *A broadband ytterbium-doped tunable fiber laser for 3He optical pumping at 1083 nm*, Applied Physics B **78** (2004), 145.

[52] M. Leduc, S.B. Crampton, P.J. Nacher and F. Laloe, *Nuclear polarization of gaseous 3He by optical pumping*, Nuclear Sci. App. **1** (1983), 1.

[53] P.-J. Nacher, E. Courtade, M. Abboud, A. Sinatra, G. Tastevin and T. Dohnalik, *Optical pumping of helium-3 at high pressure and magnetic field*, Acta Phys. Polon. B **33** (2002), 2225. Online at http://hal.archives-ouvertes.fr/hal-00002223/

[54] M. Abboud, A. Sinatra, X. Maitre, G. Tastevin and P.-J. Nacher, *High nuclear polarization of 3He at low and high pressure by metastability exchange optical pumping at 1.5 Tesla*, Europhys. Lett. **68** (2004), 480.

[55] A. Nikiel, T. Palasz, M. Suchanek, et al., *Metastability exchange optical pumping of 3He at high pressure and high magnetic field for medical applications*, Eur. Phys. J. Special Topics **144** (2007), 255.

[56] H.H. Mc Adams, *Dynamic nuclear polarization of liquid 3He by optical pumping*, Phys. Rev. **170** (1968), 276.

[57] R. Barbe, F. Laloe and J. Brossel, *Very long nuclear 3He nuclear relaxation times at 4K using cryogenic coatings*, Phys. Rev. Lett. **34** (1975), 1488.

[58] G. Tastevin, P.-J. Nacher, L. Wiesenfeld, M. Leduc and F. Laloe, *Obtaining polarized liquid 3He from optically oriented gas*, J. Phys. (Paris) **49** (1988), 1.

[59] P.J. Nacher, G. Tastevin, M. Leduc, S.B. Crampton and F. Laloe, *Spin rotation effects and spin waves in gaseous polarized 3He*, J. Phys. Lett. (Paris) **45** (1984), L-441.

[60] M. Leduc, P.J. Nacher, D.S. Betts, J.M. Daniels, G. Tastevin and F. Laloe, *Nuclear polarization and heat conduction changes in gaseous 3He*, Europhys. Lett. **4** (1987), 59.

[61] M.E. Hayden, E. Baudin, G. Tastevin and P.-J. Nacher, *NMR time-reversal as a probe of incipient turbulent spin dynamics*, Phys. Rev. Lett. **99** (2007), 137602.

[62] R.S. Timsit, J.M. Daniels, E.I. Dennig, A.C.K. Kiang and A.D. May, *An experiment to compressed polarized 3He gas*, Bull. Am. Phys. Soc. **15** (1970), 761.

[63] J.M. Daniels, L.D. Schearer, M. Leduc and P.-J. Nacher, *Polarizing 3He nuclei with neodymium La1-x Ndx Mg Al11 O19 lasers*, J.O.S.A. B **4** (1987), 1133.

[64] G. Eckert, W. Heil, M. Meyerhoff, et al., *A dense polarized 3He target based on compression of optically pumped gas*, Nucl. Instr. Meth. Phys. Res. A **320** (1992), 53.

[65] J. Becker, J. Bermuth, M. Ebert, et al., *Interdisciplinary experiments with polarized He-3*, Nuc. Instr. Meth. Phys. Res. A **402** (1998), 327.

[66] D. Rohe, P. Bartsch, D. Baumann, et al., *Measurement of the neutron electric form factor* G_{en} *at 0.67* $(GeV/c)^2$ *via 3He→(e→,e' n)*, Phys. Rev. Lett. **83** (1999), 4257.

[67] D.S. Hussey, D.R. Rich DR, A.S. Belov, et al., *Polarized He-3 gas compression system using metastability-exchange optical pumping*, Rev. Sci. Instr. **76** (2005), 053503.

[68] J. Choukeife, X. Maitre, P.J. Nacher and G. Tastevin, *On-site production of hyperpolarised helium-3 gas for lung MRI*, Abstracts ISSN 1524-6965 (2003), 1391.

[69] K.H. Andersen, R. Chung, V. Guillard, et al., *First results from Tyrex, the new polarized-He-3 filling station at ILL*, Physica B **356** (2005), 103.

[70] E.J.R. van Beek, J. Schmiedeskamp, J.M. Wild, et al., *Hyperpolarized 3-helium MR imaging of the lungs: testing the concept of a central production facility*, Eur. Radiol. **13** (2003), 2583.

[71] M. Batz, S. Baessler, W. Heil, et al., *He-3 spin filter for neutrons*, J. Res. Natl. Inst. Stand. Technol. **110** (2005), 293.

[72] A.K. Petoukhov, K.H. Andersen, D. Jullien, et al., *Recent advances in polarised He-3 spin filters at the ILL*, Physica B **385** (2006), 1146.

[73] J. Schmiedeskamp, W. Heil, E.W. Otten, et al., *Paramagnetic relaxation of spin polarized He-3 at bare glass surfaces Part I*, Eur. Phys. J. D **38** (2006), 427.

[74] A. Deninger, W. Heil, E.W. Otten, et al., *Paramagnetic relaxation of spin polarized He-3 at coated glass walls Part II*, Eur. Phys. J. D **38** (2006), 439.

[75] J. Schmiedeskamp, H.J. Elmers, W. Heil, et al., *Relaxation of spin polarized He-3 by magnetized ferromagnetic contaminants Part III*, Eur. Phys. J. D **38** (2006), 445.

[76] B. Saam, W. Happer and H. Middleton, *Nuclear relaxation of 3He in the presence of O2*, Phys. Rev. A **52** (1995), 862.

[77] L. de Rochefort, A. Vignaud, X. Maitre, et al., *Influence of lung filling on T_2^* values in human at 1.5 T with hyperpolarised ^3He*, Abstracts ISSN 1545–4436 (2004), 2724.

[78] E. Durand, G. Guillot, L. Darrasse, et al., *CPMG measurements and ultrafast imaging in human lungs with hyperpolarized helium-3 at low field (0.1 T)*, Magn. Reson. Med. **47** (2002), 75.

[79] C.P. Bidinosti, J. Choukeife, G. Tastevin, A. Vignaud and P.-J. Nacher, *MRI of the lung using hyperpolarized He-3 at very low magnetic field (3 mT)*, Magn. Reson. Mater. Phy. **16** (2004), 255.

[80] C.P. Bidinosti, J. Choukeife, P.-J. Nacher and G. Tastevin, *In-vivo NMR of hyperpolarized 3He in the human lung at very low magnetic fields*, J. Magn. Reson. **162** (2003), 122.

[81] R.W. Mair, M.I. Hrovat, S. Patz, et al., *Orientation-dependent 3He lung imaging in an open access, very-low-field human MRI system*, Magn. Reson. Med. **53** (2005), 745.

[82] A.M. Oros and N.J. Shah, *Hyperpolarized xenon in NMR and MRI*, Phys. Med. Biol. **49** (2004), R105.

[83] cf. http://www.phil.ens.fr. The PHIL project has been supported by the EC contract QLG1-2000-01559 running from 12/2000 to 06/2004.

[84] cf. http://www.phelinet.eu. Phelinet is an EC-funded Research and Training Network (RTN) of the 6th Framework Program (2007-2010). It involves 11 academic laboratories in 7 countries and 6 industrial partners.

[85] T. Stavngaard, L. Vejby Sogaard, J. Mortensen, et al., *Hyperpolarised 3He MRI and $81^{m}Kr$ SPECT in chronic obstructive pulmonary disease*, Eur. J. Nucl. Med. Mol. Imaging **32** (2005), 448.

[86] J.M. Wild, M.N.J. Paley, L. Kasuboski, et al., *Dynamic radial projection MRI of inhaled hyperpolarized 3He gas*, Magn. Reson. Med. **49** (2003), 991.

[87] E.J.R van Beek, J.M. Wild, H.U. Kauczor, et al., *Functional MRI of the lung using hyperpolarized 3-helium gas*, J. Magn. Reson. Imaging, **20** (2004), 540.

[88] F. Lehmann, B. Eberle, K. Markstaller, et al., *Ein Auswerteprogramm zur Quantitativen Analyse von Messungen des Alveolären Sauerstoffpartialdrucks (paO2) mit der Sauerstoffsensitiven 3He-MR-Tomographie*, Fortschr Roentgenstr **176** (2004), 1390.

[89] A.J. Deninger, B. Eberle, J. Bermuth, et al., *Assessment of a single-acquisition imaging sequence for oxygen-sensitive 3He-MRI*, Magn. Reson. Med. **47** (2002), 105114.

[90] M. Salerno, E.E. de Lange, T.A. Altes, et al., *Emphysema: hyperpolarized helium 3 diffusion MR Imaging of the lungs compared with spirometric indexes - Initial experience*, Radiology **222** (2002), 252.

[91] T.A. Altes, J. Mata, E.E. de Lange, J.R. Brookeman and J.P. Mugler III, *Assessment of lung development using hyperpolarized helium-3 diffusion MR Imaging*, J. Magn. Reson. Imaging **24** (2006), 1277.

[92] T.A. Altes, J. Mata, D.K. Froh, A. Paget-Brown, E.E. de Lange and J.P. Mugler, *Abnormalities of lung structure in children with bronchopulmonary dysplasia as assessed by diffusion of hyperpolarized helium-3 MRI*, Proc. Intl. Soc. Mag. Reson. Med. **14** (2006), 86.

[93] I.R. Young, *Significant events in the development of MRI*, J. Magn. Reson. Imaging, **19** (2004), 525.

Pierre-Jean Nacher
Laboratoire Kastler Brossel
ENS, UPMC and CNRS
24, rue Lhomond
F–75005 Paris
France

Progress in Mathematical Physics (PMP)

Edited by
Anne Boutet de Monvel, Université Paris VII Denis Diderot, France
Gerald Kaiser, Center for Signals and Waves, Austin, TX, USA

BIRKHÄUSER

Progress in Mathematical Physics is a book series encompassing all areas of theoretical and mathematical physics. It is intended for mathematicians, physicists, and other scientists, as well as graduate students in the above related areas. This distinguished collection of books includes authored monographs and textbooks, the latter primarily at the senior undergraduate and graduate levels. Edited collections of articles on important research developments or expositions of particular subject areas may also be included.

PMP 55: Duplantier, B. / Raimond, J.-M. / Rivasseau, V. (eds.)
The Spin. Poincaré Seminar 2007 (2009). ISBN 978-3-7643-8794-5

PMP 54: de Oliveira, C.R.
Intermediate Spectral Theory and Quantum Dynamics (2009). ISBN 978-3-7643-8794-5

PMP 53: Damour, T. / Duplantier, B. / Rivasseau, V. (eds.)
Quantum Spaces. Poincaré Seminar 2007 (2007). ISBN 978-3-7643-8521-7

PMP 52: Damour, T. / Duplantier, B. / Rivasseau, V. (eds.) Gravitation and Experiment. Poincaré Seminar 2006 (2007). ISBN 978-0-8176-8523-1

PMP 51: Mathúna, D.Ó.
Integrable Systems in Celestial Mechanics (2008). ISBN 978-0-8176-4096-5

PMP 50: Anglès, P.
Conformal Groups in Geometry and Spin Structures (2007). ISBN 978-0-8176-3512-1

PMP 49: Palmer, J.
Planar Ising Correlations (2007). ISBN 978-0-8176-4248-8

PMP 48: Duplantier, B. / Raimond, J.-M. / Rivasseau, V. (eds.)
Quantum Decoherence. Poincaré Seminar 2005 (2006). ISBN 978-3-7643-7807-3

PMP 47: Damour, T. / Darrigol, O. / Duplantier, B. / Rivasseau, V. (eds.)
Einstein, 1905–2005. Poincaré Seminar 2005 (2006). ISBN 978-3-7643-7435-8

PMP 46: Marchenko, V.A. / Khruslov, E.Y.
Homogenization of Partial Differential Equations (2006). ISBN 978-0-8176-4351-5

PMP 45: Douçout, B. / Duplantier, B. / Pasquier, V. / Rivasseau, V. (eds.)
The Quantum Hall Effect. Poincaré Seminar 2004 (2005). ISBN 978-3-7643-7300-9

PMP 44: Demuth, M. / Krishna, M.
Determining Spectra in Quantum Theory (2005) ISBN 978-0-8176-4366-9

PMP 43: Bermúdez de Castro, A.
Continuum Thermomechanics (2005) ISBN 978-3-7643-7265-1

PMP 42: Iliev, B.Z.
Handbook of Normal Frames and Coordinates (2006). ISBN 978-3-7643-7618-5

PMP 41: Cercignani, C.
Slow Rarefied Flows. Theory and Application to Micro-Electro-Mechanical Systems (2006) ISBN 978-3-7643-7534-8

PMP 40: Friedman, Y.
Physical Applications of Homogeneous Balls (2004). ISBN 978-0-8176-3339-4

PMP 39: Colombo, F. / Sabadini, I. / Sommen, F. / Struppa, D.C.
Analysis of Dirac Systems and Computational Algebra (2004). ISBN 978-0-8176-4255-6

PMP 38: Dalibard, J. / Duplantier, B. / Rivasseau, V. (eds.)
Poincaré Seminar 2003. Bose-Einstein Condensation – Entropy (2004) ISBN 978-3-7643-7106-7

PMP 37: Nikiforov, A.F. / Novikov, V. / Uvarov, V.B.
Quantum-Statistical Models of Hot Dense Matters. Methods for Computation Opacity and Equation of State (2004). ISBN 978-3-7643-2183-3